DISCOVERING THE
HUMAN BODY

DISCOVERING THE HUMAN BODY

How pioneers of medicine solved the mysteries
of the body's structure and function

BERNARD KNIGHT, M.D.

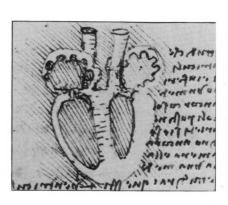

Lippincott & Crowell, Publishers
New York

Discovering the Human Body
by Bernard Knight
was conceived, designed, and produced
by Imprint Books Limited
12 Sutton Row, London W1V 5FH

Art Direction: Susan Rose
Designer: Fiona Almeleh
Editorial: Paulette Pratt
         Ellen Sarewitz
Production: Kenneth Clark
          John Reynolds

FIRST EDITION

U.S. Library of Congress Cataloging
in Publication Data

Knight, Bernard
  Discovering the human body

  Includes index.
  1. Human physiology—History. 2. Anatomy,
Human—History. 3. Body, Human. I. Title.
QP21.K57   612'.009   80-7886
ISBN 0-690-01928-9

Printed and bound
by TONSA, San Sebastian
D. L. SS 333/1980

Contents

Foreword

For centuries, stories of famous explorers have been immensely popular and the names of such men as Christopher Columbus and Vasco da Gama have been familiar to generations of readers. A recent phenomenon is the increasing interest of laypeople in explorers of a much more intimate world—that which is inside their own bodies.

The last decade has seen an almost explosive interest in human biology and medicine—men and women suddenly seem to have discovered themselves. In a climate of increasing health education, there is a new awareness of the ways in which the body can be abused by modern living, and as medical knowledge has more and more been extended to large sections of the public, popular expositions of medical subjects have proliferated. The days of a patronizing professional elitism have gone and a public acutely aware of its "right to know" has torn down the barricades that formerly protected professional knowledge.

This book reveals one vital aspect of that knowledge—the historical background of the way the structure and function of the human body was discovered. It is vital because in any subject no true appreciation and no real perspectives can be gained unless the historical basis is established. The story of the discovery of the human body has two faces. One is the straightforward recording of the advances made by certain men at certain times. The other is the beguiling practice of eponymy, the commemoration of a new discovery by attaching to it the name of the discoverer.

This fascinating cult of eponymy has had its rise and its partial decline. The ancients rarely linked their names to body structures, but as dissection and description accelerated through the Middle Ages and the Renaissance, eponyms began to appear in profusion. Then, in the last part of the nineteenth century, a reaction set in and official censure of anatomical eponyms was enforced. Happily, it is often ignored and most anatomists still retain their own private store of famous names, even if it is not on the same scale as in the clinical subjects or in such other sciences as geology or zoology where, without eponyms, there would be little left to learn.

The knowledge of what lies beneath the human skin and how it works has been painfully and slowly built up over two and a half thousand years. In the pages that follow, we have traced this evolution of knowledge as it went—slowly, then faster, then stopped and even reversed for about a millennium, before the headlong acceleration of the past century brought us to the present state. Anatomy has yielded just about all its secrets, although physiology, the science of how the body works, has many mysteries yet to be solved.

The rise of eponyms during the scientific Renaissance was largely the result of increasing academic intercourse between men in different parts of Europe. All using Latin, the *lingua franca* of scholarship, they exchanged ideas and facts in personal letters and by the publication of books. In the fifteenth to seventeenth centuries, there were few journals and periodicals, such as now carry the bulk of scientific discoveries. Medical men had to indulge in long and frequent correspondence with each other and with the few learned societies that existed. Anton van Leeuwenhoek, the Father of Microscopy, for example, became known to the world almost solely through the hundreds of letters he wrote to the Royal Society of London. In this close world of personal relationships, it was natural that new discoveries should be credited—and sometimes wrongly—to a person well known by name to his academic brethren. It was both a compliment and a convenience to have the arteries of the brain named after Thomas Willis or a brain fissure after Sylvius. In addition to such immediate eponyms, the classically educated polymaths of the Renaissance tended to produce attributions to Greek mythology and the Achilles tendon and the Atlas vertebra are eponyms of a different type.

Thus anatomy became built up of a mass of eponymous tissues and organs, although the discipline could not compare in this respect with other sciences which consisted almost totally of proper names moulded into Latin or Greek forms. Indeed, geology, zoology, and botany were solidly founded on eponyms and remain so until this day.

Towards the end of the nineteenth century, some revolt began against anatomical eponyms. Strangely enough, one of the greatest rebels was Friedrich Henle, who himself was commemorated in a whole string of eponyms. In 1887, a meeting of the Anatomical Society in Leipzig failed to agree on any action, but at the 1895 Congress at Basel, it was decided to use the plain anatomical name first, with the eponym following parenthetically. This so-called Basel nomenclature survived in textbooks until 1933, when the Birmingham Revision dispensed with the bracketed names. Finally, in 1955 the Paris Congress decided that in gross anatomy eponyms should not be used at all.

Many have regretted—and ignored—this call to bury the names of the famous anatomists of the past. In spite of the dictates of Paris and Birmingham, they still use many of the better known names. It is a rare doctor who does not think of the Fallopian tubes, the Eustachian tubes, or the circle of Willis. Their colleagues in other branches of medicine, particularly neurology and pathology, still use eponyms avidly—and in many cases find them most convenient, especially when a disease is insufficiently understood to make the application of a coldly descriptive name truly accurate.

Anatomical eponyms breathe life into the subject. Such names as Wirsung's duct can conjure up a memory of a Bavarian doctor assassinated in Padua in the seventeenth century, an image which the sombre pancreatic duct can never create. Similarly, the premaxillary bone hardly stirs the emotions, but the knowledge that it is also named after Goethe, Germany's greatest poet—who fought a duel at the age of ten—must surely inject some interest into all but the most unimaginative of minds.

In this book, the history of how things were discovered, how they got their names, and after whom they were named, is set out system by system, with other more general discussion of the broader sweep of anatomical and physiological history, such as the story of dissection itself, the very basis of anatomical science. Apart from eponymy, there is fascination to be found in anatomical etymology, the origins of the actual words applied to parts of the body. These pages reveal an extraordinary spectrum of such names, taken from the classical languages, ranging from Turkish saddles to bridal chambers and from reed pens to French combs.

But these curiosities are merely the icing on the cake. What really matters is the saga of discovery over several thousand years—sometimes laborious, sometimes brilliant. Often men were prejudiced, envious, and even malicious in their dealings with their colleagues. The reactionary attitudes of religion stunted progress for centuries and led to oppression and martyrdom. The dogma of the ancients warped men's attitudes for more than a thousand years and the giant spectre of Galen inhibited progress from Roman times until a few centuries ago.

The story of the discovery of the human body is by no means finished, especially in the realm of physiology. But as an interim report of the first few millennia, this account must be unique.

Bernard Knight.

Introduction

Throughout the ages, hundreds of men—and one or two women—have given their names to parts of the human body, but far more have contributed anonymously to our understanding of the structure and function of the most marvellous piece of machinery ever created.

We have no idea who the first anatomists were, and we can never know nor hope to find out, for every person is, to some extent, an anatomist. Everyone studies the human body. Some may admire and respect its dimensions and proportions; others may revel in the aesthetic delights of its supple mobility. Some appreciate the ripple of strong muscles under the skin or stir to the sensual line of thigh or breast.

Earliest people must have looked at their own bodies and those of their fellows with increasing awareness and understanding. Like so many other things, the importance of a structure and its functions becomes greater when it goes wrong—illness and injury are great concentrators of the mind. Inevitably, in the meanest of early caves and dwellings, men and women must have been puzzled by a broken arm or a paralyzed leg and must have sought ways to restore the limb to usefulness.

Not only human anatomy would have come under scrutiny. The very meat they ate, from rabbit to bison, must have increasingly aroused interest as human perception sharpened. Bones, tendons, the meat of the muscles, the strangely shaped organs—all these would have provoked curiosity. Later, in some cultures, organs began to assume magical qualities of prophecy and divination. The Babylonians, for example, used liver models for soothsaying, and until the Christian era entrails were "read" in Celtic countries.

Through early observations of the bodies of animals and people, the crude foundations were laid of a science that was to evolve, albeit with many blind alleys and dormant periods, over thousands of years. That ancient people knew at least the rudiments of animal and even human anatomy has been amply proved by archaeological discoveries. Cro-Magnon men and women cared for their sick as best they could; their main medicine was magic. The artists of that era left clear evidence in the form of small statues, carvings, and cave paintings. Perhaps the earliest example is the Willendorf Venus, a tiny limestone statuette from the Aurignacian period,

The Willendorf Venus from the Aurignacian period, about twenty thousand years ago, is one of the earliest examples of an artist's concept of anatomy.

about twenty thousand years ago. The figure appears grossly obese, with huge thighs, abdomen, and breasts suggesting some glandular disorder. But it is probable that the carved model represents an ancient concept of the ideal woman or, more likely, that it was a fertility object, thus explaining the accentuated sexual characteristics.

Of more direct anatomical interest is a reindeer horn found in the great cavern of Mas d'Azil in the Ariège Department of southern France. It was carved in Magdalenian times, some twelve thousand years ago. Three horses' heads were incised into the horn with exquisite skill; the first shows the surface features, the next has been skinned to reveal the contours of the muscles, and the last is virtually a bare skull. All this points to a dissecting skill, even if only in the course of butchery, and an appreciation of the layered structure of the animal's body.

Cave paintings of the same early Stone Age period record elephant and bison, some with the outline of the heart quite accurately positioned on the animal. Interestingly, the shape of the heart depicted all those millennia ago is exactly the same as appears today on playing cards and Valentines, with the pointed apex and the two rounded humps of the auricles. Other paintings have arrows lying in the area of the heart, where they would most effectively kill the beasts. It is not clear whether these were sympathetic magic to ensure good hunting or were training murals for spearmen and archers, but the shape and position of the vital organs reveal some anatomical knowledge as well as an understanding of which structures are the most vulnerable. These paleolithic anatomists were also crude physiologists.

Although much is known about the clinical methods and therapy of the great civilizations of the Near East, India, and China, there is only fragmentary, and tantalizing, evidence of any prowess in anatomy. The Egyptians, for example, who should have had the greatest knowledge of anatomy, left relatively few records of their expertise, and those that exist tend to be bizarre. Although they mummified at least seventy million of their dead after elaborate evisceration, only fragmentary and disappointing evidence of their knowledge of the body has survived. This may be because a special class of priest, not doctors, performed the last rites.

The civilization of ancient Mesopotamia had quite elaborate medical services, but little is known of any anatomical knowledge, other than what is mentioned in descriptions of surgical procedures. In Babylon, the liver was first postulated as the seat of the essential life process, a concept that held sway until much later when the primacy of that organ was displaced by the heart. The liver played an important role, too, in the Babylonian practice of divination diagnosis. A sick patient would breathe into the nostrils of a sheep, which was then sacrificed so that its liver could be examined. A diagnosis was reached by comparing accurate clay models of the liver (some of which have survived until today) with the liver of the sacrificed sheep. Much later the Etruscans practised the same method of diagnosis using models of bronze.

India also had its ancient anatomical traditions, although it is difficult to establish a chronology. One of the most famous writers was Susruta, who lived in the fifth century A.D., but he quoted much ancient material, some of which went back to about 700 B.C. Among the many branches of medicine described, surgery was prominent, but it seems that anatomical knowledge was fanciful rather than factual. The nerves and blood vessels were thought to radiate from the navel, and it was evident that numbers had some intrinsic fascination for the doctors; it was alleged that there were three hundred bones, ninety tendons, two hundred and ten joints, five hundred muscles, seventy blood vessels, three humours, three kinds of secretion, and nine sense organs. Blood vessels were thought to carry air.

Ancient Indian medical texts recommended that every doctor should dissect a body for himself, although from their anatomical check-list it does not seem that the process produced much practical benefit. Since the Hindu laws of the time prohibited the use of a knife when dissecting, a corpse was immersed in water for a week, so that the putrified body could be examined merely by pulling the various parts asunder.

The great civilization of China was isolated from the mainstream of medical progress until relatively recent times, although some of their concepts percolated into Western medicine. The West's theory of the four humours of the body, for example, was basically a variation of the Chinese concept of a balance between the Yin and the Yang. In ancient Chinese anatomy, the heart was considered the most important organ, with the liver its mother, the stomach its son, and the kidney its enemy. Each organ was matched with a colour, a plant, and a season, with elaborate symbolism attached to the viscera and their functions. The two principles of life, the male and female Yang and Yin, were believed to fuse together with the blood to form the "vital spirit" which circulated through the body fifty times each day. All diseases were thought to be due to disharmony and imbalance between the Yang and the Yin.

These anatomical concepts were fanciful, based on a theoretical idea of the interior of the body, rather than on dissection. Dissection, however, usually on the bodies of condemned criminals, may have been practised as long ago as 1000 B.C., since it is mentioned in the *Ling Shu*, an ancient Chinese book of anatomy.

The Greeks, of all the ancient civilizations, were unassailably the leaders in medicine and anatomy—as indeed they led in so many other things from sculpture to democratic government. "The Greeks have a word for it," the old saying claims, and, indeed, the very word "anatomy" was coined by them. It means dissection and derives from the Greek words for cutting up.

The earlier centuries of Greek history can be disregarded as far as anatomy is concerned. It was only in the post-Homeric period, starting about five hundred years before the birth of Christ, that the first recognizable attempts were made to understand the body and its workings.

The pre-eminence of the Greeks in laying the foundations not only of anatomy, but of all medicine and science, was quite possibly the result of their geographical position at the crossroads of Asia, the Middle East, North Africa, and Europe. Most of their famous medical men came either from the Aegean islands or from Greek Asia Minor, rather than from the mainland. Nevertheless, it is necessary to keep going back to the Greeks at every turn in medical history, at least until the Renaissance when western Europe began to build in earnest upon the Greek foundations.

Even if a few seeds of knowledge were imported from the East, most Greek medical discoveries and theories were purely home-grown. Medicine was not only the province of the doctors, but of such philosopher-scientists as Aristotle, Plato, and Empedocles.

The actual beginning may well have been the work of Alcmaeon, who lived in the Greek colony of Croton in southern Italy in the fifth century B.C. He dissected animals and recognized a multitude of structures, including the optic nerve and the canal leading from the throat to the middle ear, which was later named for Eustachius, the sixteenth-century Italian anatomist.

In Sicily, Empedocles, a contemporary of Alcmaeon, began to develop some theories of bodily function. It was he who first said, "Blood is life." He attributed the life force to the "innate heat" of the body, which he believed was distributed by the heart. Here arose the concept of "pneuma"—an insubstantial substance that was both soul and life—which permeated the blood vessels. More than just a constituent of the human body, pneuma was believed to be the

This clay object, which resembles a sheep's liver and is inscribed with magical formulae, was used for purposes of divination by the priests of Babylon in about 2100 B.C.

9

Discovering the Human Body

Empedocles, who lived in the fifth century B.C., was a major Greek philosopher. He came from Agrigento in Sicily, where he is said to have been offered a kingship. Little is known about his life—or his death, although legend has it that he slipped away from a feast in his honour and threw himself into the crater of Etna so that no trace would be found of his going and it would be presumed that he had been taken by the gods. He is credited with introducing the four elements as the basis of the system of humoral pathology which affected medical theory and practise for two thousand years.

The Hippocratic Oath, sworn in the names of various gods of ancient Greece, was originally laid down as an ethical statement for the healing family of the Aesculapiads. The message contained in the oath, however, which is one of integrity and vocational guidance, has survived almost two and a half millennia, to be sworn still by newly qualified doctors all over the world.

all-pervading life force of the universe and could be seen ephemerally as a shimmering cloud when the blood of a sacrifice was shed. Although the concept of pneuma had been voiced before—by Anaximenes, for example, half a century earlier —Empedocles was regarded as the founder of the pneumatic school of medicine, which was to survive in one form or another for many centuries. Like the Chinese, who had also regarded the heart as the centre of life, the pneumatists were concerned with the philosophical concepts of man in relation to the universe, the microcosm and the macrocosm.

In the late fifth century B.C., two opposing schools arose—the followers of Empedocles and the other of dissenters. Diogenes of Appollonia, a confirmed pneumatist, described the vascular system in not too extraordinary a fashion, except that he made the common error of placing twin blood vessels coursing through the trunk of the body, rather than the single large aorta. The rivals were the Hippocratean School of Cos.

Hippocrates himself, the great clinician who is known as the Father of Medicine, probably wrote little of what is attributed to him in the so-called Hippocratic Collection. This might be just as well, as far as his reputation as an anatomist is concerned, for although something was known of the bones and skull, much of the rest was speculative gibberish, obviously owing nothing to dissection. Nerves, for example, were not recognized as such, although the term itself was applied to tendons and sinews. The brain was thought to be a gland that exuded mucus, and the auricles of the heart were believed to be receptacles for air.

Hippocrates's son-in-law, Polybus, was a somewhat better anatomist. He was a recluse who cut himself off from the world and its pleasures to devote himself to the study of the body. He wrote two treatises, one *On Man* and the other *Nature of the Child.* But his concept of anatomy was crude and inaccurate. He stated, for example, that the main blood vessels consisted of four pairs, which ran by the most extraordinary and unlikely routes from one part of the body to another; the heart did not seem to figure much in this theoretical scheme. It was obvious that he had never seen the inside of a human body.

The purely speculative works of Plato contributed nothing to anatomical fact, but his philosophical concepts, such as those offered in the *Timaeus,* influenced people until the Middle Ages. Here the doctrine of the macrocosm of the outer world of the universe was compared with the microcosm of the human body, yet this did less than nothing to assist the difficult birth of genuine anatomy.

Aristotle was the first Greek to get to grips with the realities in any substantial way. The son of a physician, but not a doctor himself, he laid the foundations for an understanding of evolution and was, indeed, the founder of comparative anatomy. He was also an ardent naturalist, and it is clear that his descriptions were the result of extensive dissection of animals. He corrected the erroneous descriptions of his predecessors about the great blood vessels and was the first to use the term aorta.

It was Aristotle who first described and named numerous structures. The earliest known anatomical drawing is Aristotle's diagram of the male uro-genital system. He drew up the *Scala Naturae,* which was a kind of evolutionary ladder diagram. In embryology, his descriptions and drawings were remarkably accurate, especially his work on the dogfish placenta. It was obvious, however, that all his dissections were on animals, and that he never had the opportunity to examine the structure of the human body. But for his time, his research was outstanding and his influence can be clearly discerned through the Middle Ages and later.

It must be admitted, of course, that Aristotle's errors were quite considerable. He never attributed any importance to the brain, for example, and he placed the seat of life and intelligence firmly in the heart. The brain he believed to be merely an organ for cooling the heart, by preventing overheating. This cooling function, Aristotle claimed, was carried out by the secretion of "phlegm." (It is interesting to note that the pea-sized endocrine gland at the base of the brain is today still called the pituitary, its name being derived from the Greek *pituita,* or phlegm.) He also believed that the arteries contained air, not blood, a common belief in those times; again, the word artery derives from "air."

Aristotle developed the system of the four humours, which had been first clearly described by Polybus, to embrace a comprehensive arrangement in which the two biles, the phlegm, and the blood, were combined with fire, water, earth, and air, as well as heat, cold, dryness, and wetness. All natural phenomena could be brought within some permutation of this complex equation, including sickness and health. Only in comparatively recent times has this system, devised in the fourth century B.C., fallen into disuse.

Despite his misconceptions, Aristotle is a giant in the history of medicine, not only because of his descriptive research, but also because of his philosophical ideas, especially in relation to the nature of generation. His basic concept of life depended not on whether substances were organic or inorganic, but on whether they had *psyche*—a rough equivalent to "soul." This was not merely a theological argument; it permeated his theories of reproduction, and of life and death. He believed that the female contributed the material substance of the embryo via the egg, but that the male gave the principle of life, insemination being the breathing of the *psyche*. Yet Aristotle considered the passage of seminal fluid only incidental to this process. The essential contribution of the male was not matter but principle, and this led him to consider the possibility of parthenogenesis. From this, the allowability of virgin birth was taken up in later centuries by the Christian Church, which thus approved of many of the pagan Aristotelian theories.

When Greece fell to Rome, the torch of medical learning was handed to Alexandria across the Mediterranean. The great academic centre on the coast of Egypt became the refuge of classical scholarship for several centuries. The first of the Alexandrian anatomists was Herophilus who was said to be the first man to dissect both animal and human bodies, though this surely must have referred to public demonstrations. In refutation of Aristotle's opinion, Herophilus restored the brain to its rightful place as the seat of intelligence. He knew, too, that nerves conduct sensation and direct muscle movements. Many anatomical structures and names are to be credited to him, and one, a blood vessel in the back of the skull, commemorates his name eponymously. He was the first to make a clear distinction between arteries and veins, but thought that pulsation in the arteries was due to their own active movement, rather than to a pressure wave running through them. He named many other structures, especially in the brain.

Herophilus probably dissected more human bodies than anyone before his time, gaining first-hand knowledge of what the organs and tissues actually looked like. Later, he and his rival and contemporary, Erasistratus, were accused of human vivisection. The allegations, some made

Hippocrates, the Father of Medicine, was said to have descended from the healing god Aesculapius and became a legend even in ancient Greece. He created the art and science of medicine and removed it from the realm of superstition and magic. Born on the island of Cos in the Aegean in 460 B.C., he learned medicine from his father, a physician, and probably studied philosophy with Democritus. Hippocrates lived to an advanced age and travelled widely; he may even have visited Alexandria. The oldest representation of him was found in the remains of an ancient temple of Aesculapius on Cos, dating from the fourth century B.C. Tradition identifies a spreading plane tree growing in the centre of the town of Cos as the one under whose branches Hippocrates received his patients. There, too, it is said he taught his pupils his holistic theory of the relationship between doctor, patient, and disease in the diagnosis and treatment of the sick. At the time this was a departure from traditional therapy, for illnesses were believed to be inflicted by the gods as a sign of their displeasure. Cures were equally mystical. Hippocrates, however, listened to his patients, rather than to the oracles. His rational approach to healing did not survive even in the classical world. Only recently has its full significance been grasped and begun to be embodied in modern medicine.

Aristotle, the Greek philosopher and the Father of Biological Science, was born in Stagira in Macedonia in 384 B.C. His father was court physician to Philip, the King of Macedon. Aristotle was sent to Plato's Academy in Athens in 367 where he remained until his teacher's death. He then travelled for twelve years, establishing two new academies, and went back to his home where he tutored the young Alexander. In 355, Aristotle returned to Athens where he opened his Lyceum, a rival to the Academy. There, he created a centre for speculative research into almost every field of study. Until his death in 322 B.C., he made outstanding contributions in embryology, biology, comparative anatomy, political science, and philosophy. As a natural scientist he has had few peers. Perhaps his most important contribution was his systematic method of enquiry and the ordering of acquired knowledge. Yet he was primarily a philosopher and for him science was subservient to philosophy. He preferred logical

deduction to observation, an approach which gave rise to gross errors in his own research. Aristotle's work first came to western Europe indirectly, through the Arab philosophers. Whereas he advocated freedom of thought so that people could search for the truth, the universities that adopted Aristotelianism demanded rigid adherence to what they considered to be given truth. Consequently, the errors contained in his scientific works became not the starting point for further research, but the enshrined, if incorrect, statement of biological truth.

Erasistratus of Chios, born in about 310 B.C., was said to have been the grandson of Aristotle. Unlike Aristotle, who saw the soul as innate in being, he adhered to the atomistic philosophy of Democritus, and saw nature as an external force imposing life on the body. In contrast to the holistic methods of Hippocrates, he applied specific therapy to specific complaints. This preoccupation with the function of individual organs earned him the title Father of Physiology. His extensive studies of the circulatory system were the result of his efforts to discover where in the body air was transformed into vital spirit and animal spirit.

Herophilus was born about 300 B.C. and was a contemporary of Erasistratus at Alexandria. Little is known of his life and no portrait of him exists. All his writings have been lost, but he seems to have studied the brain in particular. In contradiction to Aristotle, he identified the brain, rather than the heart, as the seat of intelligence. He had a reputation as an innovator. He conducted the first public demonstrations in anatomy and he is said to have taught the first woman medical student at Alexandria. He was vilified by early Christian theologians in subsequent centuries for practising human vivisection, but there is no evidence to substantiate such accusations.

Aristotle was possessed by a desire to learn the truth of nature and Christian theologians adopted his philosophical methods to express instead the truth of faith. Although a pagan, he was regarded as the patron saint of the scholars at centres of learning in Europe, including Chartres, where his figure appears on the cathedral's façade.

by St. Augustine, were probably false, but Herophilus was said to have dissected some six hundred persons and to have caused the death of foetuses by opening living wombs.

Erasistratus was more of a physiologist than an anatomist, and he and Herophilus made a complementary pair in the Ptolemaic Medical School in Alexandria. Erasistratus recognized that a tripartite bundle of artery, vein, and nerve, goes to every organ and part of the body, and he invented a highly complex theory to justify his observations in terms of physiology. The blood, according to Erasistratus, was carried by the veins, but air was taken in by the lungs to be changed into a type of pneuma called "vital spirit," which passed into the heart and was distributed to the body by the arteries. When this spirit reached the brain, it was changed into "animal spirit," which was then dispersed by the nerves. In some things, particularly related to the heart and circulation, Erasistratus came nearer to some of the truth than others managed for another eighteen hundred years.

Alexandria survived into the first part of the Christian era. Eventually it declined, especially in medical science, and the greatest academic calamities in history were the destructive attacks on the great university library, first by fanatical Christians and later by the Arabs. To trace the story of medicine and anatomy the Greeks must be followed back to Europe, to Rome itself.

Even in Rome, at the hub of the Empire, almost all doctors were Greeks, since the Romans considered medicine a menial occupation fit only for slaves and immigrants. There was a medical school in Rome from about 60 B.C. founded by Asclepiades of Bythynia. Anatomy was disregarded and from the early years of the Christian era human dissection was forbidden—and remained so until the late Middle Ages. There was undoubtedly surgical prowess in the Roman system of medicine, especially in the military services, but this was not generally

founded on any great anatomical knowledge.

Although the Greeks had almost a closed shop as far as medicine was concerned, one notable exception in the field of scientific writing was Celsus; there are hints, however, that much of his work was founded on older Greek texts. A Roman nobleman of the family of the Cornelii, Aulus Cornelius Celsus lived at about the time of Christ. He was not a doctor himself—that would have been an occupation beneath the dignity of a Roman patrician—but he had a profound interest in all aspects of contemporary life, from agriculture to military tactics.

Celsus wrote a great encyclopaedia, dealing with these subjects as well as with law, medicine, and philosophy. Most of this work has been lost, but eight books dealing with medicine survived and were discovered fourteen centuries later. These were entitled *De Re Medica* and were republished at Florence in 1478, on the orders of Pope Nicholas V. They were among the first medical books to be printed with movable type, the new process that was to revolutionize human thought and endeavour. Written in fluent Latin —which was unusual, since most other medical books were in Greek—the last two books of *De Re Medica* were devoted to surgery and included a considerable amount of practical anatomy, including a complete description of the skeleton.

Celsus knew the details of the windpipe and gullet, the diaphragm and lungs. He was unaware of the duodenum, already described by Herophilus, and had only a fair idea of the surgical anatomy of the bladder and the female genital organs. But it is his knowledge of the bones and the skull that is remarkable, considering that he was not even a physician and would have had no opportunity to examine a dead body. Particularly good were his descriptions of the bones of the arm and leg, presumably gained from discussions with military surgeons, for his knowledge of battlefield surgery was excellent. The practicality of his approach, the clarity of his writing, and the almost modern surgical techniques he described, make Celsus one of the world's outstanding figures in medical literature.

In the Roman period, there is another Greek from Asia Minor to whom deference is due. Although his personal character and standard of writing were far inferior to those of Celsus, Claudius Galen nevertheless deserves the title Prince of Physicians. Also known as the Dictator of Medicine, Galen's views on medical science and the philosophy of medicine were so completely accepted that after his death, in about A.D. 201, men stopped doing medical research for more than one thousand years, since it was believed that Galen had discovered all there was to know, and further work was therefore futile.

Galen's grip on the minds of learned men was so complete that it was considered heresy to dispute his conclusions. Even as late as 1559, the Royal College of Physicians of London made one of their members, Dr. John Geynes, retract his statement that there were some errors in the writing of Galen. Geynes hastily complied.

Galen wrote more than three hundred books, of which parts of one hundred and eighteen have survived. His books—the errors included—became the main repository of medical knowledge through the Dark Ages. Galen held a long and successful position at the seat of power in Rome, despite many brushes with professional colleagues early in his ascension, because of his aggressive, domineering manner.

Although his clinical interests ranged broadly, in the context of anatomy Galen essentially consolidated and enlarged upon the current knowledge of the day. Most perfect was his description of the skeleton, especially the skull and spine. His muscle anatomy was less impressive, but better than any up until that time. With the exception of his great and, no doubt, excusable errors concerning the movement of the blood, much of his descriptive writing was a great advance on anything that had gone before. He was the first, for example, to declare that the arteries contain blood, not air and, what is more, he proved it experimentally. He was not able to practise dissection, although he had examined the skeleton, but he traced the main arteries quite accurately. In general, however, his writing about the cardio-vascular system is the least impressive.

When it came to the nervous system, Galen stated that the brain was the centre of the nerves of sensation and the spinal cord was the centre for the control of motion. He described only seven cranial nerves out of the twelve, but his accounts of many of the brain structures were excellent, considering they were made from animal understudies. Many anatomical terms still in daily use, especially in the bones and the

Claudius Galen, the son of Nicon, a prominent mathematician and architect, was born in Pergamon in Asia Minor in A.D. 129. When he was fifteen, his father sent him to study philosophy, but the next year Nicon decided, as the result of a dream, that his son should have a career in medicine and Galen began to attend the temple of Aesculapius to study the people who came to be healed by the semi-religious, semi-medical methods of his time. A doctor by the age of twenty-one, Galen's first book was *On the Anatomy of the Uterus,* and within a few years he had written works on diseases of the eye and the chest. He undertook further studies at the medical school at Alexandria, then came home when he was twenty-eight. For four years he held the post of medical officer to the gladiators. Doubtless this presented him with a great deal of human anatomy, but the morality of the day forbade human autopsies, so his dissections were limited to the beasts brought out from the games. Like all aspiring young men of the time, Galen finally went to Rome. Ambition as well as ability took him to the top of his profession. He aroused the jealousy of other doctors, but earned the favour and trust of the emperor, Marcus Aurelius. Galen went with the Imperial Army on its campaign to Germany, and returned to Rome with the emperor. Before his next expedition, Marcus Aurelius entrusted his son Commodus to Galen's care. Galen used this as an excuse to remain in Rome and write more books. His output as a writer was prolific and it was said that he kept twelve scribes continuously occupied. Many of his writings were, however, destroyed by a fire in his home. Galen died in about A.D. 201. Although he was a great clinician, the philosophy and arrogance which directed his discoveries and writings crippled the progress of medicine for the next fifteen hundred years.

nervous system, were donated by Galen. His contributions were great, but errors abounded and it was the later prohibition of any criticism of Galen's misapprehensions that caused so much trouble centuries afterwards. Anatomists steeped in Galenic-Aristotelian traditions would not—or could not—bring themselves to deny the hallowed words of the long-dead Greeks, even when their own eyes perceived the falsity of the more fantastic theories.

Galen's physiological theories were even more fault-ridden than his anatomical descriptions. He drew upon the philosophy of pneuma to try to explain the workings of the body. Following the general precepts of Aristotle, he tried to rationalize the reasons for a particular structure, in terms of the functions it was expected to perform, in the context of humours and spirits.

He developed a complex and devious explanation of heart and breathing function, compounded from several earlier writers. He declared that the basic principle of life was drawn in from the general world spirit, or pneuma. It entered the trachea during breathing and passed to the lungs from where it passed through the pulmonary vein to the left ventricle; there it first encountered the blood. The blood was derived from the intestinal tract, with the liver converting it into venous blood imbued with "natural spirit." This ebbed and flowed through the venous system, part of which reached the lung through the pulmonary artery and discharged its impurities. Another part filtered through tiny pores in the septum, or wall, between the left and right ventricles. When it reached the left side of the heart, it came into contact with the pneuma, which had entered through the trachea, and the contact produced the "vital spirit"; this was then distributed through the arterial system. Because some of these arteries went to the brain, the vital spirit reached that mysterious organ and was changed into yet a third type of spirit, the "animal spirit"; this was distributed through the nerves, which were said to be hollow.

Galen was a pagan in a pagan land, but his theories had strong religious overtones and he mentioned both Jewish and Christian beliefs in his writings, although it was obvious that he had no personal leanings towards either. He justified the cause and effect of a God-made part of a body which was miraculously fashioned to be right for the job it was expected to carry out. This is no doubt the reason the Christian Church later accepted Galen so readily—and jealously protected him against all dissenters, to the virtual exclusion of progress for the next fifteen hundred years. Consequently, his utterly fantastic and complex description of the heart and breathing functions remained in vogue until the seventeenth century and could not be displaced by any more rational ideas until that time.

The discoveries and theories of Claudius Galen were in many ways a last glimmer before the Dark Ages plunged medicine and science into an eclipse that was to last until the Renaissance. With the fall of Rome, the Greek physicians dispersed, and soon western Europe was in the grip of a tyrannical Church that was totally averse to any scientific progress.

This detail from the frontispiece of a seventeenth-century collection of the Hippocratic writings shows four medical greats—Galen, the Prince of Physicians, Hippocrates, the Father of Medicine, Avicenna, the Persian physician whose *Canon* was the most famous medical book ever written, and Aetius, the sixth-century Greek physician who compiled an extensive medical work, the source of valuable historical material.

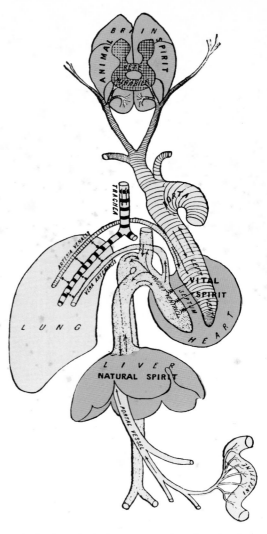

Galen's physiological system was based on the three pneumas—the natural spirit, the vital spirit, and the animal spirit.

A great deal of medical knowledge was irretrievably lost—many of the only copies of Galen's works were destroyed in a fire during his lifetime, and the burning of the books in Alexandria was a catastrophe on an unprecedented scale. It was the Islamic world that saved much of Western scholarship from the ruins of the Roman Empire and from the disinterest or outright opposition of the Christian Church.

A series of doctors, some Arab, some Jewish, obtained the surviving works of Aristotle, Hippocrates, and Galen, and kept them alive in Arabic translation in the Moorish world that had extended across the Middle East and North Africa as far as Spain, in the incredible expansion of the eighth century. Some of the books written by these doctors, such as that of Avicenna, remained standard textbooks through the Middle Ages. Eventually, they were retranslated into Latin, the lingua franca of academics until recent times. The original Greek became available to

western Europe through the alien tongue of Islam; without this repository, anatomy, and medicine generally, might have lost the progress of centuries.

In southern Europe, a few centres of learning managed to survive the collapse of the Roman Empire. The most notable example was Salerno in Italy, and, not far away, at Monte Cassino, the translation of the Arabic works was begun in about 1050 by a monk, Constantine.

Anatomy, along with all of medical science, was at its lowest ebb in the tenth and eleventh centuries. No dissections could take place since they were abhorred by the Church. In fact, none had taken place since the years before Galen, who may have, on one solitary occasion, examined the body of a drowned woman.

The Christian Church taught that this life was merely a chafing delay before the ascent into a greater life after death. Consequently, trivialities like anatomical science were irrelevant. Death was an obsession, and the medieval men and women thought always of their demise. Little wonder that, together with the denial of experimental science, no progress was made in these centuries. Indeed, far from progress, much of what was known from ancient times became distorted and devalued.

Mystical interests grew, especially in astrology, and the Zodiacal Man became a common illustration, a magical attempt to relate bodily parts and illnesses to the movement of the heavenly bodies. Although these beliefs were of pagan origin, the Church did not seem to proscribe them, but the attitude was that people should concentrate on their souls, rather than their bodies.

In the twelfth and thirteenth centuries, there was a gradual awakening, with a resurgence of learning in the new universities of Italy, then France, and later Britain and the Germanic countries. Anatomy, in the scientific sense, began in these new institutions, haltingly at first, as they were still under the steely eye and iron hand of the Church, which later used the Inquisition as a powerful deterrent to heresy of any kind. Yet, academics became progressively more adventurous, until the virtual explosion of experimental science that began in the sixteenth century.

For anatomy, this rebirth was largely the result of a practical need, rather than academic searching. This was the necessity of postmortem examinations for legal purposes; thus forensic medicine can claim a longer pedigree than anatomy, at least in the new world of the scientific method.

Probably the birth of modern anatomy can be traced to Bologna, where the first true university was founded. It began as a law school, but a medical faculty, the first in Europe, was estab-

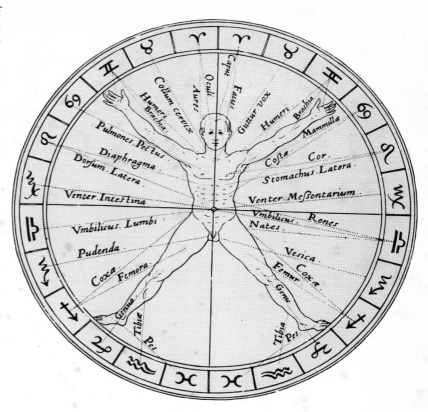

lished there as early as 1156. Teaching consisted of readings from the re-translations into Latin of the Greek works which had been compiled in Arabic by Avicenna, perhaps the greatest of the Muslim medical men. But the lawyers of Bologna wanted medico-legal opinions on dead bodies, and from these forensic autopsies the opportunity arose to discover anatomical as well as legal truths. The demand for such autopsies became great at about the end of the thirteenth century, and, before long, they developed into anatomical dissections.

The Bologna medico-legal autopsies were a way of evading a Church ban on all dissection. Pope Boniface VIII had issued a Papal Bull in 1300 prohibiting the cutting up of dead bodies. The actual reason for this law was the practice prevalent among the Crusaders of defleshing the bodies of their mortally wounded comrades and boiling the skeleton, so that the bones could be sent home, thus avoiding burial in a heathen land. When the Church banned this, it was interpreted by zealous priests as a prohibition of all dissections.

Bologna remained true to its pioneering form and appointed a first professor of anatomy, Mondino de Luzzi, who was active in the early years of the fourteenth century. He was the first of the Western-style anatomy teachers, and though his teaching consisted solely of readings from Galen, it was at least a beginning.

One of his pupils was Guy de Chauliac, himself to become a Bolognese disciple in Paris. He

In the Dark Ages, when scientific progress was at a standstill, interest in astrology grew and for hundreds of years belief persisted in magical zodiacal influences on the organs of the body.

described Mondino's teachings: "My master, having placed the body on a table, gave four discourses over it, the first dealing with the nutritive organs, because they putrify first. . . . In this way, he teaches anatomy from the bodies of men and animals and not from pictures."

Mondino himself earned the title of the Restorer of Anatomy, and his textbook, the *Anathomia*, published in 1316, was the first to be devoted solely to anatomy. It was not a great book; it was written in bad Latin with confused nomenclature, and much of it was taken from the Arabic texts. Nevertheless, it was a step forward and, given the limitation of Mondino's anatomical material, it was a worthy venture.

Bodies were hard to come by in those days, and usually the only opportunities for dissection other than the medico-legal post-mortems, were on the remains of executed criminals. Because of the speed of decay, dissections had to be done in haste, the procedure often continuing day and night until it was finished. The criminals were almost all male, so little opportunity arose to examine women.

The first definite reference to a dissection was in 1286, when a plague was ravaging northern Italy. A physician of Cremona opened a body to try to find the cause of the pestilence, but it seems that he only looked into the chest to view the heart. In 1301, a case of poisoning was investigated in Bologna, and a post-mortem was carried out by two physicians and three surgeons under the direction of the professor of medicine, Bartolomeo da Varignana. The autopsy report ends with the words, "we have assured ourselves by the anatomization of the parts." By the fourteenth century, dissection was well established in Italy, but northern Europe was more conservative, and body-snatching from gallows and graves was relatively common. An illustration of an early autopsy in Britain depicts a monk and a physician watching a more lowly surgeon removing the organs from a body.

The early writers, including Mondino, slavishly followed the old Galenic and Aristotelian dictates preserved in the Moorish literature—and they preferred it to the testimony of their own eyes. If the old theories and the actual evidence differed, as they frequently did, then it was assumed that the body undergoing dissection was abnormal. Nobody ventured to disagree with the book.

For the next few hundred years, anatomical dissections followed a formal, stereotyped pattern. The august professor would mount his podium and sit high on his ornate chair, like a bishop on his throne. (This is the origin of the "chair" of a university, the actual seat of the eminent professor.) At anatomical lectures, the professor would read out passages from the textbooks while an assistant stood below and,

with a baton, pointed out the structures as they were mentioned. The assistant was called the ostensor and he never actually touched the body himself. The dissection was done by a demonstrator, a menial. The students stood around the body, watching and listening, but their eyes were always subservient to their ears. The whole process was a reaffirmation of Galen's autocratic views and it was only later, when the professor came down from his exalted chair and began to look for himself, that any real progress was made.

Late in the fourteenth century, attitudes began to change—and continued to change with increasing speed as the new century dawned. The restraints imposed by the Church began to be loosened, but indulging in progressive thought was still a risky business, and at least one anatomist went up in flames along with his books.

Artists as well as scientists began to see man in his natural place in the great scheme of things, instead of as a formalized figure, intent only on his soul and the great hereafter. Painters and sculptors began to represent the body as it really is; gone were the stiff puppets of religious art of past centuries.

The sculptors and artists were better surface anatomists than the men of science, and the rise

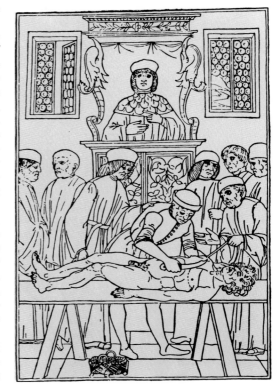

This fifteenth-century woodcut of an anatomy lesson at Padua shows the professor in his chair, reading from a book, while a menial demonstrator dissects, as directed by the wand of the ostensor. The students in academic dress take no active part in the dissection.

of naturalism in art led them to take an intense interest in the human form. Lucio Signorelli, Andrea Mantegna, and Andrea Verrochio were followed by Michelangelo, Dürer, Raphael, and Leonardo da Vinci—all of whom made detailed studies of the human body.

Only in relatively recent years has the great anatomical genius of Leonardo come to light. He took art into anatomy. To vitalize his drawing and painting, he began to study the body. But his fascination with anatomy soon took him deeper into the subject. He began to do dissections, as did many of his artist contemporaries, and this developed into his absorption with the structure and the function of the whole body. His drawings, famous not only for their beauty and accuracy, but also for the odd mirror-image handwriting, were lost until recent years, and, therefore, had little effect on the mainstream of anatomy during the Renaissance. Leonardo had intended to write a textbook in collaboration with Marcantonio della Torre, a professor at Padua, but the death of the academic prevented this. Had it come to fruition, such a book might have considerably accelerated anatomical and medical knowledge.

After Mondino, there was a succession of mediocre men at Bologna, but when printing with movable type began in the late fifteenth century, a number of books by Italian professors of anatomy and surgery appeared. The first was printed in Venice in 1493. In 1521, while teaching at Bologna, Jacob of Berengar brought out a commentary on the work of Mondino; it was the first such book to be illustrated. Berengar wrote the worst Latin ever published, but he ventured to deny some of Galen's statements, including the existence of a network of vessels at the base of the brain, which does not occur in man. He was also the first to describe the appendix, the thymus gland, and parts of the brain.

Other books followed, including those by Johannes Dryander of Marburg in Germany and Charles Estienne of Paris, and anatomy was now no longer an exclusively Italian preserve. Students from Bologna, Padua, and other Italian universities had gone home, and schools were springing up in many places in addition to the more ancient universities, in Montpellier, Paris, and Basel.

By the early sixteenth century, the University of Padua had taken precedence over Bologna as the centre of medical scholarship in Italy. It was here that the greatest anatomist of that time taught. This was Andreas Vesalius, son of a Belgian pharmacist. Whereas Mondino had been called the Restorer of Anatomy; Vesalius earned the title the Reformer of Anatomy; it was he who finally came down from the professorial chair and instituted observation in place of dogma and speculation.

Today, with several hundred years of the

The frontispiece of the 1493 edition of *Anathomia* by Mondino de Luzzi depicts an anatomical demonstration before the master. The demonstrator is probably Alessandra Giliani, the brilliant girl who assisted Mondino with his dissections.

Leonardo da Vinci was the epitome of the Renaissance man. His genius is still unrivalled. The illegitimate son of a Florentine lawyer, he was born in 1452. He was apprenticed to the artist Verrochio and became a master in the painters' guild at the age of twenty. Ten years later he entered the service of the Duke of Milan, to whom he was defence advisor, architect, mathematician, anatomist, botanist, designer, and painter. Although his artistic achievements were prodigious, Leonardo spent most of his time engineering military works for his patron. In 1536, he went to the court of King Francis I of France. Until his death three years later, he spent more time on mathematics and hydraulics than on painting. One of the greatest biological investigators of all time, Leonardo dissected more than thirty bodies and made many fine anatomical drawings, which are said to have been the basis of the illustrations in Vesalius's *De Fabrica*.

scientific method behind us, this approach would seem so obvious as to make the alternative ridiculous, but it is Vesalius and others like him who deserve the credit for initiating the modern way of thinking. Until his time—and, indeed, for some time afterwards—learning consisted of agreeing sycophantically with the dictates of the old masters. Even Vesalius himself and many of his successors were so steeped in this attitude, that they were unable to shake off entirely the looming shadows of the early giants.

Vesalius denied the existence of the tiny pores in the septum of the heart which had been avowed by Galen. Even the great Leonardo, although he was unable to see the pores himself, had agreed that they must exist if Galen had said so. Vesalius, not without considerable courage, denied the existence of the pores. It must be admitted, however, that in the first edition of his great work, *De Humani Corporis Fabrica* (On the Fabric of the Human Body), he went along with the Roman master and only in his second edition did he feel confident enough to offer a refutation.

The work was first published in 1543, after a mere three years of hard work at Padua, but in that short time Vesalius set medicine and, indeed, much of experimental science, on a new road. By declaring that "seeing is believing" he had started a revolution in man's approach to science.

After Vesalius left, Padua continued to be justly famous in the realm of anatomy. He was succeeded by men like Matteus Realdus Columbus, the discoverer, or rather re-discoverer, of the lesser circulation through the lungs, which was vital to the later realization of the greater circulation. The same discovery had been made independently by Michael Servetus six years before, but poor Servetus perished in the flames of a Calvinistic inquisition.

Other famous men at Padua included Fallopius, whose name is remembered in the uterine tubes, Cesalpino, who had a tenuous claim to anticipating Harvey's discovery of the circulation, Fabricius, Eustachius, and others who, although they did not have the eminence of Andreas Vesalius, added lustre to the brilliance of Italian anatomy. The details of their contributions will appear throughout this book, for they either left their names attached to the structures they described, or had some direct part in important discoveries. Vesalius's name was linked, however, only to a few trivial anatomical structures, some of which were abnormalities. Yet his great monuments were his book and his reputation, which have established him as the most important anatomist of all time.

From the sixteenth century onwards, great advances were made in the understanding of anatomy, and in medicine generally. There were still many reverses and culs-de-sac, but in the seventeenth century much of the newly gained information was consolidated. Yet it was in that century that there was a single discovery, which, like Vesalius's term at Padua, was a landmark not only in the factual sense, but in the broader philosophy of science.

The publication in 1628 of William Harvey's discovery of the circulation of the blood was research *par excellence*, a great example of the application of observation, experiment, and logical, deductive thinking in the pursuit of scientific progress. Harvey's discovery was a beacon light in scientific methodology.

Andreas Vesalius exploded the foundations of medieval medicine and roused the fury of the scientific world. Appointed professor of anatomy at Padua when he was twenty-three, he accomplished his life's work in five years and retired at the age of twenty-nine. Vesalius was born in Brussels in 1514. His father's family had been physicians and pharmacists to the Hapsburg court, and he showed an interest in medicine at an early age. His education began at Louvain where he was imbued with the ideals of the Renaissance. For his medical training he went to Paris where the teachers still adhered firmly to the principles of medieval scholarship.

Taught by ardent Galenists, Vesalius became disgusted with the imprecise method of autopsy according to the accepted texts, regardless of what the prosector discovered in dissection. He began to collect his own anatomical material by rifling graves for bones, by stealing corpses of hanged criminals, even storing bodies in his rooms for a fortnight at a time. In this way, he gained a thorough, first-hand knowledge of human anatomy. By the time he took up his first post at Louvain, he had shown his exceptional gifts as an anatomist. In 1536, Vesalius went to Padua for his doctorate. He took his degree within the year and was promptly given the chair by the Senate of Venice.

His first books were translations of Arabic texts, but he soon found them inadequate and began to write his own. In 1543, he completed his monumental *De Humanis Corporis Fabrica* which was published in Basel. Superbly illustrated, it contradicted many of Galen's theories and did not show any pores in the heart. It was greeted with outrage.

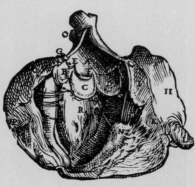

Vesalius relinquished his chair and became physician to the Spanish court. He made his last journey, to Jerusalem, in 1564. Because the Church held anatomical studies in extreme fear, it may have imposed a penance on him for expiation of some sin, now unknown. He died en route to the Holy Land.

THEATRVM ANATÓMICVM Lycei Patauini

The famous anatomy theatre at the medical school in Padua was designed and built by Fabricius of Aquapendente, the Italian anatomist who taught, and strongly influenced, William Harvey.

For centuries, even millennia, anatomical progress had been compounded of a blend of incomplete observation, religious belief, extrapolation from animal structures and philosophical guesswork, based on what a doctor thought the truth should be, rather than on what he could demonstrate. Paradoxically, William Harvey, the London physician who overthrew this approach, was a conservative, pious man, steeped in respect for Galen and Aristotle. Yet it was he, albeit unwittingly, who drove the largest nails in the coffin of Galenic influence.

It seems inconceivable today that any intelligent man, let alone a physician, could think otherwise than that the blood circulates around the body. It has been suggested, however, that until the mechanical pump was invented for mining purposes in the late Middle Ages, the concept of a pumping heart would not have arisen, since there was no model that the mind could grasp.

The influence of Harvey's discovery widened to encompass all fields of medical endeavour. Thereafter, medical men thought in terms of physiological function when they were engaged in anatomical study. The discovery of better technology, notably Leeuwenhoek's microscope, gave the opportunity and impetus for an almost geometrical progression in the pace of medical knowledge. For, although Harvey discovered the circulation of the blood, he had no idea how it got from the arteries to the veins to complete the circuit. He had no means of knowing. It was left to the Italian anatomist Marcello Malpighi many years later to discover the capillary vessels, using the primitive but adequate microscope of his time.

In the centuries that followed, discovery followed upon discovery, each one leading to a series of others, so that progress became fan-shaped, rather than linear as it had been in earlier years. By this time, however, anatomy was no longer the fastest-moving of the medical sciences, but refinement of detail and accuracy had become all-important. The rectification of the mistakes of the old masters was a major task in itself and the application of the sum total of anatomical knowledge to surgery and medicine was the practical outcome of the more academic exercises.

In later years, system after system of the body was subjected to minute dissection and microscopy. Every organ gave up its structural secrets and these were applied to physiological investigation of function where that was possible, although in this area there is still much that is obscure, particularly in the nervous and immune systems.

Today progress tends to be step-wise, rather than a smooth line on a graph; when a new technique is introduced, it is applied wholesale to all the old problems. Just as light microscopy opened a whole new world, so did immunology, endocrinology, and electron microscopy.

As the function became inseparable from the structure, anatomists came closer to physiologists. Not only did the structure of the adult body become almost totally described, but the whole mysterious process of generation and procreation was opened up to science by a cavalcade of investigators from the seventeenth century onwards. Embryology became a sub-discipline of anatomy and just as anatomy was the partner of surgery, so embryology became the colleague of obstetrics and gynaecology.

In the nineteenth century, specialization began. No longer could one man be a total anatomist, except on the level of undergraduate teaching. If he conducted research, then he had to settle on one, or at the most a handful, of localized interests. The amount of material to be read and learned was so vast that both the academics and the surgeons (who were often the same people) of necessity became particularly concerned with the eye or the ear, or the heart, or the organs of generation.

This is the story of all those men, as well as of the earlier pioneers who, in spite of their shortcomings in modern terms, were but the products of their age and who did their best within the confines of the times in which they lived.

Structure and Framework

This crude representation of a skeleton, a symbol of Epicureanism with a wine-skin extended in each hand, is from a Pompeiian mosaic of the first century A.D.

The human body is a complex mechanism which receives information and responds to that information with movement and speech. All the functions of the body are ultimately subservient to the needs and well being of the brain. The complex servo-mechanisms of circulation, respiration, digestion, and excretion, together with the activity of the nervous system, are parts of the machinery which provides energy and communication for the outward manifestations of human life.

To house both the brain and its servo-mechanisms there is an integrated framework composed mostly of bone and fibrous tissue. Sheathing the skeleton is the muscle mass, the largest single component of the body. This in turn is covered by more fibrous tissue and skin.

The general shape of the human skeleton is familiar enough, but there is a tendency to think of the bones as fundamentally distinct from the soft tissues. Most people visualize the skeleton as an isolated structure, with all the other tissues

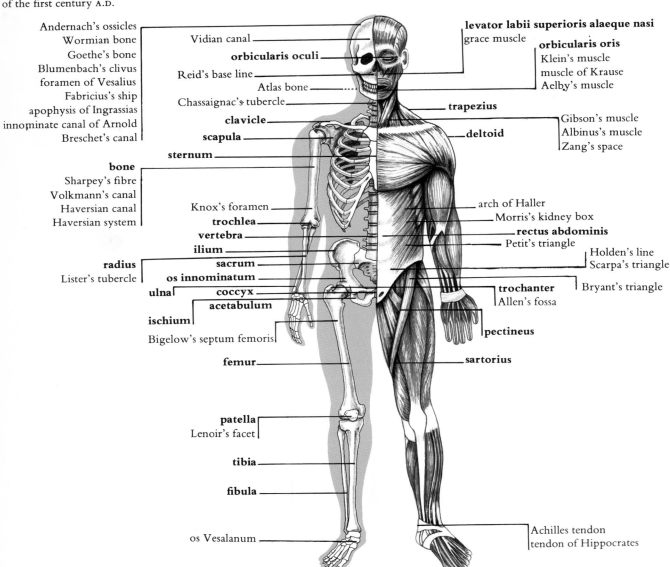

Andernach's ossicles
Wormian bone
Goethe's bone
Blumenbach's clivus
foramen of Vesalius
Fabricius's ship
apophysis of Ingrassias
innominate canal of Arnold
Breschet's canal

bone
Sharpey's fibre
Volkmann's canal
Haversian canal
Haversian system

radius
Lister's tubercle

ulna

ischium
Bigelow's septum femoris

Vidian canal
orbicularis oculi
Reid's base line
Atlas bone
Chassaignac's tubercle
clavicle
scapula
sternum

Knox's foramen
trochlea
vertebra
ilium
sacrum
os innominatum
coccyx
acetabulum

femur

patella
Lenoir's facet

tibia

fibula

os Vesalanum

levator labii superioris alaeque nasi
grace muscle
orbicularis oris
Klein's muscle
muscle of Krause
Aelby's muscle

trapezius

deltoid

Gibson's muscle
Albinus's muscle
Zang's space

arch of Haller
Morris's kidney box
rectus abdominis
Petit's triangle

Holden's line
Scarpa's triangle

Bryant's triangle

trochanter
Allen's fossa

pectineus

sartorius

Achilles tendon
tendon of Hippocrates

draped about it, like a suit of clothes on a coat-hanger or a brick building erected around a framework of steel girders. This is a convenient concept, especially for teachers of anatomy, but it is the reverse of the truth.

It was the soft tissues, not the skeleton, which came first. Both in evolution and in the developing embryo the skeleton emerged as a dense thickening in response to the need for support. Consequently, a leg should not be thought of as a core of bones with flesh attached. Rather it is a mass of muscles with specific tasks to perform, which of mechanical necessity had to develop a hard centre, appropriately jointed, to provide mobility and to compete successfully with the forces of gravity and acceleration.

This concept of the body's framework came late in the history of anatomy, particularly since many centuries passed before the theory of evolution was accepted, or even contemplated. During the hundreds of years of the Church's tyranny, questioning the immutability of the God-made body would certainly have been considered heresy.

Evidence that the shape of the bones is dependent on function can be seen when injury or disease leads to a one-sided defect. A person who has one lame leg experiences modification of the bones of the other leg—to accommodate the changes in stress and load-bearing capability. Even if the external form of the bones remains the same, changes occur in the fine latticework inside the bone, and these can be seen on X-ray. This is an example of Wolff's law—that all changes in the function of bones are attended by definite alterations in their internal structure—which was established in the late nineteenth century by Julius Wolff, a professor of orthopaedic surgery in Berlin.

The framework of the body consists, then, of everything except the specific organ systems and the liquid tissues of blood and lymph. What is left are the bones, the muscles, the fibrous tissue, and the skin.

The origins of osteology, the study of the skeleton, are unknown. Bones would have been more convenient and less unpleasant objects to handle than the intact corpse, however, and it is reasonable to assume that the general shape of the human skeleton was well understood from early times. There is a part of the Hippocratic Collection of writings which is entitled *On the Nature of Bones,* but this is misleading since it deals almost entirely with veins. Some of the more clinical parts of the Hippocratic writings mention skeletal anatomy in passing, such as that of the shoulder in dislocation, but they are far from systematic. The Alexandrians, although they dissected extensively, added little osteological description.

In about the first century A.D., the Roman

Because of artistic conventions in the Middle Ages, figures were flat, lifeless, and anatomically imprecise. This engraving, a portrayal of a skeletal anatomy lesson, appeared in a manuscript of 1363 by Guy de Chauliac, one of the most renowned surgeons of his time.

writer Celsus contributed considerably to knowledge of the skeleton. Much of his writing was orientated towards surgical treatment, military surgery in particular, but he gave a good account of many bones, especially of the limbs. That a fair knowledge of osteology existed in Roman times is evidenced by the little metal skeletons, often made of silver, which were placed on banqueting tables, or brought in with the wine, to remind guests of their mortality.

Among the hundreds of books written by Galen a century later, there was one called *Bones for Beginners*, and the descriptions in this were quite good. Many of the words still in use for various bones were introduced by Galen, but not all those usually attributed to him were actually his.

Julius Pollux, a little-known Roman and a contemporary of Galen, introduced a large number of anatomical words in his book *Onomasticon*, which means vocabulary. Dedicated to the Emperor Commodus (a patient of Galen's), the book was lost until 1502, but from that date it became a source for the Latin and Greek words used to displace many of the old Arabic terms. It was Pollux who first used, for example, the word trochanter, meaning a bony knob—like the one at the top of the thigh—to which muscles are attached.

Pollux's source for anatomical terms, however, was a small work on anatomy, titled *The Names of Different Parts of the Human Body* which was written by the Roman physician Rufus of Ephesus, who lived about A.D. 100. Rufus was said to have written more than one hundred books, but very few have survived.

Galen classified bones into those having a hollow marrow cavity and those, like the flat bones of the skull, which do not. He seemed to be aware of most of the complex bones of the skull, and recognized that there are twenty-four vertebrae in the spine. It was Galen who gave the name coccyx to the rudimentary human tail. The derivation is from the Greek word for cuckoo, because of the bone's resemblance to the bird's beak. Galen also described the sacrum, the block of fused vertebrae at the base of the spine which forms the back part of the pelvic girdle. But he wrote in Greek, and later Latin translators mistook his meaning of sacrum. While he meant to convey that it is the most important bone of the spine, it appeared in translation as the "sacred bone."

In the millennium which followed the fall of the Roman Empire, there would not have been the same overtones of sacrilege in the examination of dry bones as there was in anatomical dissection. Nevertheless, detailed knowledge of the skeleton remained poor until the Middle Ages.

It was Leonardo da Vinci who, in the fifteenth century, drew the first adequate picture of bones (although his work, lost until recent times, would not have been available to those who followed him). His drawings were faultless. He was the first to do perspective sketches, depicting bones from different angles, and he even used

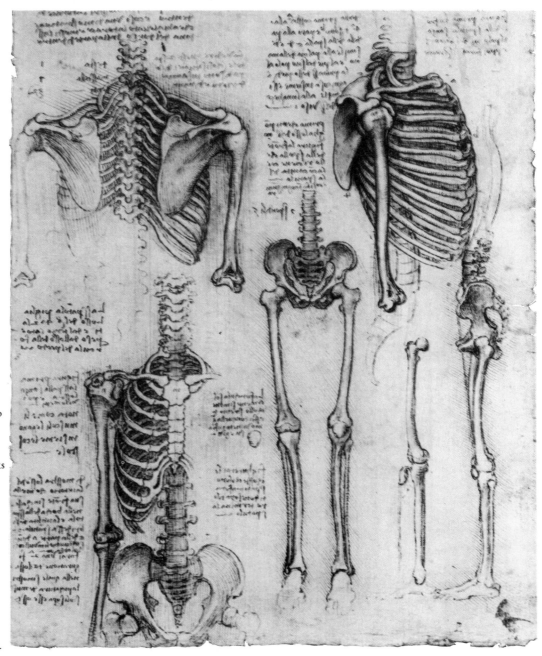

Unaccountably lost for two hundred years, Leonardo da Vinci's anatomical drawings were among a bound volume of notebooks which came into the possession of the English crown in the seventeenth century. Leonardo gained his great knowledge of anatomy from dissections he undertook himself. The anatomical notebooks contain a wealth of detailed drawings, all annotated in his curious mirror script. His early studies, of skeletal anatomy, were remarkable.

the technique of showing half of the skull cut away to reveal the internal air sinuses, a convention widely used today in medical illustration. He drew the vertebrae and the spinal column with a realism that had never before been achieved and which was not to be matched again. Leonardo's drawings of the shoulder muscles, too, were far clearer and more informative than those appearing in many modern textbooks.

In the sixteenth century, Italian anatomists began to publish their books on anatomy. The pinnacle of achievement among these early printed anatomy books was Vesalius's great work *De Humani Corporis Fabrica*. In the same year, 1543, he also published an alternative version, the *Epitome*, for readers who were not students of medicine. With these two books, Andreas Vesalius, the founder of modern observational anatomy, produced the first comprehensive account and illustrations of the human skeleton, as well as most of the other systems.

Although by no means without error—the influence of the long-dead Galen was still pervasive—*De Fabrica* and *Epitome* included superb illustrations of the body surface and the skeleton. Drawn by the Flemish artist Jan Stefan van Kalkar, the figures were depicted in life-like poses, with the muscles shown in a sequence of progressive dissection to illustrate the layered structure. Vesalius's anatomy was living and the drawings were active and mobile.

The first of the seven volumes of *De Fabrica* was devoted to the bones and joints. It followed the general pattern of Galen in the classification of the skeletal structures, although here and there Vesalius had the temerity to contradict the great man. He denied, for example, that humans have a separate bone in the front part of the upper jaw, the pre-maxillary bone, which exists in certain animals. He included a special illustration of a human skull lying across that of a dog, to emphasize the point. Up until this time, too, it was solemnly believed that the male had one less rib than the female because of the description in *Genesis* of Adam's loss. Strangely, however, in what is probably one of the most famous anatomical drawings in existence—that of a skeleton leaning pensively against a tomb contemplating a skull—the hyoid bone, the little, horseshoe-shaped structure in the neck, is that of a dog. Like many anatomists before him, Vesalius, or the artist van Kalkar, confused, or unwisely transposed, animal structures with those of man.

In the skeleton, Vesalius himself is commemorated in two places, but they have proved variants of the normal pattern. In the illustrations in his books the so-called os Vesalianum or bone of Vesalius, regularly appears. It is a small spherical bone lying between the base of the little toe and the cuboid bone in the instep. But this bone exists only in rare cases where part of

the base of the inner bone of the little toe has separated. Similarly, the central bone of the skull, the sphenoid, was first described by Vesalius. Among the many pits and tunnels in this complex bone, there is occasionally an extra hole, which allows the passage of a small vein. This gap, which is only rarely present, is called the foramen of Vesalius. It is ironic that the name of one of the greatest anatomists of all time only occurs eponymically in relation to two structures which are merely sporadic curiosities.

Andreas Vesalius needed no eponyms, however, to secure his niche in the history of medicine, for his influence went beyond the realms of

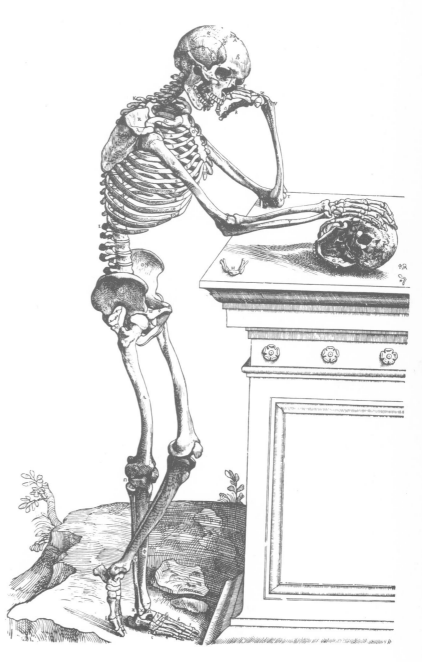

The pensive skeleton leaning on a tomb is one of the most famous illustrations in the first volume of Vesalius's *De Fabrica*, which was devoted to bones. Next to the detached skull are the hyoid bone— mistakenly that of a dog— and two of the ossicles of the ear.

anatomy and science. He was the epitome of the academic and scientific Renaissance man.

It is perhaps appropriate that the oldest anatomical presentation in existence is a skeleton put together by Vesalius for a lecture he gave at the University of Basel in Switzerland, the city where *De Fabrica* was published. Concluding his demonstration, which took place in 1546, three years after his monumental work appeared, Vesalius presented the skeleton to the university, where it remains.

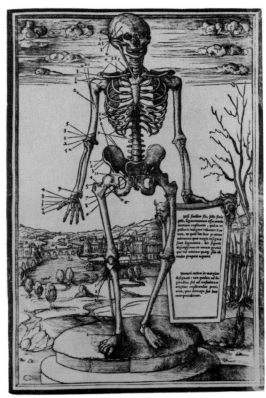

This skeleton appears in *De Dissectione Partium Corporis Humani*, written by the French physician Charles Estienne. Begun in the pre-Vesalian era, it was not published until 1545—two years after *De Fabrica*. The main defect of Estienne's work was the ugliness of its figures.

The names of most of the bones are derived from the ancient Greek or Latin, although in many cases they were first applied by the Renaissance anatomists, rather than by the ancients in their time. Some names have suffered in translation. Many original Greek words were distorted by the poor classical scholarship of the Middle Ages, and an added complication was that some had to be translated second-hand from Arabic versions of the original Greek.

The seams, or sutures, which run between the skull plates on the top of the skull are called, for example, sagittal, coronal, and lambdoid sutures. The sagittal suture, running fore and aft, comes from the Latin for an arrow; the coronal suture across the head is from corona, or crown; and the forked lambdoid suture at the back of the head is named for its similarity in shape to the Greek letter *lambda*. The large, smooth surface of the upper-arm bone, which meets the elbow, is called the trochlea, meaning a pulley—an apt

functional description. The word pelvis comes from the Latin for a bowl, due to the deep concavity of the pelvic girdle. Inspiration seems to have run out with the two halves of the pelvis, however, since each of these is referred to as an innominate bone (os innominatum), being unnamed. The large, cup-shaped socket of the hip joint is known as the acetabulum, the vinegar bowl. In the shin, the main bone is the tibia, from the Latin for a flute, and its long, thin partner, the fibula, means a clasp or brooch.

A considerable number of men left their names attached to some part of the skeleton. Many of these are, however, relatively insignificant or inconstant structures.

Of the more important eponymous structures the Wormian bones in the skull, which are a congenital variant, are of more interest to the forensic pathologist than to the anatomist, since they are sometimes mistaken for bullet wounds.

Most often occurring at the back of the head, Wormian bones are the result not of disease or malformation, but of developmental accident. Sometimes in the foetus little areas of the cranial membrane begin to ossify away from the main centres of bone growth, along the edge of the skull plates. These areas are rapidly overtaken by the main process of ossification, but they persist into adulthood as small islands adjacent to or incorporated into the main seam, or suture, lines which join the hard plates of the skull. These little islands, the Wormian bones, are often symmetrical, and, although usually only two or four are present, up to a hundred have been recorded in rare cases.

Although these little sutural bones were known to the ancients, it was the Danish physiologist Claus Worm who first described them accurately in the seventeenth century. Claus Worm was a member of the Bartholin family of medical and scientific geniuses from Copenhagen, and it was his nephew Thomas Bartholin, the professor of mathematics and anatomy, who honoured him with the eponym. Actually,

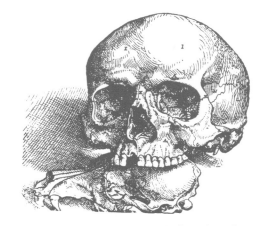

Vesalius superimposed a human skull on that of a dog to signal a mistake in Galen's description.

these little bones had been described earlier, although not so well, by an anatomist in France, Gunther of Andernach. He described in 1536 what were called Andernach's ossicles, until Worm's more accurate description. Later, in the eighteenth century, the German poet, Johann Wolfgang von Goethe described a larger Wormian bone at the lambdoid suture, which became known as Goethe's bone. This world-famous German playwright and poet, who lived from 1749 to 1832, was also a student of comparative anatomy. Goethe is also commemorated in the premaxillary bone, which is not present in the adult, but is a developmental structure at the front of the bony palate.

Blumenbach's clivus is the name given to the sloping shelf of bone which leads up the inside floor of the cranium from the foramen magnum (the hole through which the spinal cord exits from the skull) to the sella turcica, which means Turkish saddle and evocatively describes the pea-shaped cavity in the floor of the skull which holds the tiny pituitary gland. This smooth slope of bone, on which lies the vital brain stem, was named in 1786 by Johann Blumenbach, the German physiologist and anthropologist and one of the most celebrated doctors of his time in Europe. Also in this area, a small blood vessel penetrates the bone at the base of the skull. It lies in what is known as the Vidian canal, recalling the sixteenth-century Italian anatomist Guido Guidi, who was better known by the Latinized name of Vidius.

The cranium is made of plates of bone constructed like a sandwich, with a thin but dense layer on the outer and inner sides, called the tables, and with spongy bone in between, which is known as the diploe. In this diploic space run a number of venous canals, first described by Gilbert Breschet in 1819 and later called Breschet's veins and canals. A professor of anatomy at the University of Paris, Breschet has other structures named after him.

The small part of the sphenoid bone, forming part of the inner wall of the eye socket, enjoys the name the apophysis of Ingrassias. It was named for the sixteenth-century osteologist Giovanni Filipo Ingrassias who has been called the Sicilian Hippocrates. He described the bone in a book which was published in 1603, twenty-three years after his death.

Also in the sphenoid bone there is a small, inconstant tunnel which conducts the lesser petrosal nerve from the hard mass of bone which surrounds the delicate ear. In a convulsion of illogicality, this canal is called the innominate canal of Arnold, although it is difficult to understand how it can be considered unnamed. It was first described in 1834 by Friedrich Arnold, who has a string of eponyms to his credit, mainly in the nervous system. Arnold had a distinguished career, for he was professor of anatomy in turn

Johann Wolfgang von Goethe, the greatest of the German poets, made significant contributions to anatomical knowledge. Born in Frankfurt in 1749, he was sent by his father to study law. But his classical studies, and the men that he met, inspired him to a new, romantic genre. He had a number of passionate affairs and developed an interest in the occult. His work in the sciences was notable for its painstaking observations, but his theories were more the product of a philosophical view of nature than of a scientific method. Goethe died in 1832, one of the last polymaths of the Enlightenment.

Giovanni Filipo Ingrassias was born in Sicily in 1510. A physician and a professor of anatomy and medicine, first at Naples and then, from 1563, at Palermo, he specialized in the study of bone structure. He identified the descriptions in Galen's works on the skeleton as those of the bones of monkeys and not of humans. Ingrassias was also an epidemologist. During the plague of 1575–76 in Palermo, he recorded his observations. He distinguished measles from scarlatina, and was probably the first person to recognize chicken pox as a separate disease. Ingrassias realized that disease is caused and spread by an external agent rather than by an imbalance of humours, or miasmas. He died in 1580.

at the universities of Zürich, Freiburg, Tübingen, and, finally, Heidelberg.

A curious eponym in the skull is Fabricius's ship, which compares the profile of the main bones of the wall and base of the skull to the keel and sides of a ship's hull. It is rather sad that this vague and fanciful eponym is the only one in the body which recalls Fabricius of Aquapendente, one of the great line of anatomists from Padua and the man who taught and inspired William Harvey.

In the post-cranial skeleton—in other words, all the rest of the bones—one of the best-known named parts is the topmost vertebra of the spinal column, the Atlas bone. This first cervical vertebra, which supports the globe of the head, takes its name from the Greek god Atlas, whose task it was to carry the heavens on his shoulders.

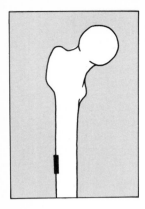

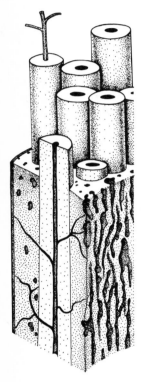

Haversian systems, which consist of densely packed concentric cylinders of bone, form the substance of compact bone. Each of these cylinders is served by its Haversian canal, carrying blood, lymph, and nerve supply to the bone.

Lower down in the neck, the little wings, known as transverse processes, on each side of the sixth cervical vertebra, both have little knobs on the front surface. These knobs, which are known as Chassaignac's tubercles, are vital in emergency surgery because the main carotid arteries can be squeezed back against them as a desperate procedure to staunch serious bleeding higher in the neck or face. This is not without danger, for sudden pressure on the carotid artery, which carries blood to the brain, can trigger a nervous reflex causing the heart to stop. This is the fatal mechanism seen in death from the karate or commando punch. These important little tubercles were described in the nineteenth century by a Parisian surgeon Charles Marie Chassaignac.

Of particular interest to chest specialists is the angle of Louis, which is the name given to the slight bend towards the top of the breastbone. This bend may become exaggerated in patients with certain chest diseases where the effort of breathing produces a "pigeon chest." It was named after Pierre Charles Alexander Louis, a Parisian chest physician, renowned for his work on pulmonary tuberculosis, who described it in 1825. Louis was one of the leading French physicians of his day. He made detailed observations of two thousand cases of chest tuberculosis in three Paris hospitals and performed almost four hundred autopsies. He made a similar accurate study of typhoid fever and was one of the first doctors to introduce statistical methods into clinical medicine.

Elsewhere in the skeleton, eponyms are rather sparse. In the arm there is Lister's tubercle on the radius, the outer bone of the forearm. It is named after Joseph Lister, the pioneer of antiseptic surgery, who described the bony nodule in 1865 in an article in the *Lancet* in relation to a surgical operation on the wrist.

The other lower arm bone, the ulna, is accommodated in a deep pit in the upper arm bone, or humerus, when the elbow joint is moved to its full extent. This pit is known as Knox's foramen, because it was described by Robert Knox in 1841, although his name is more likely to be remembered for his involvement in the notorious Burke and Hare murders.

In Edinburgh in the late 1820s, when dissection was still illegal and corpses were therefore hard to come by, William Burke and William Hare set themselves up in business to supply bodies to Dr. Knox, who was described as an "unofficial anatomist." The first one or two corpses had died natural deaths, but then these ambitious entrepeneurs turned to murder. Most of their victims were derelicts and prostitutes and the final toll was unknown.

The trial opened on December 24, 1828. Hare turned king's evidence and was, therefore, not tried. Knox was not even asked to testify. But Burke was tried, found guilty, and hanged.

A small pit on the neck of the thigh bone, known as Allen's fossa, is named after Harrison Allen who was professor of anatomy at the University of Pennsylvania and president of the Association of American Anatomists in the later part of the nineteenth century. On the inside of the femoral neck, there is an arched pattern of ridges called trabeculae, from the Latin for little beams, which are strengthening girders to support the great mechanical stresses that occur in the neck of the thigh bone during walking, running, and jumping. One of these, much thicker than the rest, is called Bigelow's septum femoris. It was described by another American, Henry Jacob Bigelow, in 1869, in relation to the common fracture of the neck of the thigh bone. Bigelow was professor of surgery at Harvard from 1849 to 1882.

The knee cap, or patella, is the largest sesamoid bone (meaning similar in shape to a sesame seed) in the body. It is a bony thickening in the massive tendon that throws the lower leg forwards from the knee joint in walking or kicking. Gliding over the lower end of the femur, it allows the tendon to change direction around the corner of the knee when the leg is bent. Underneath, the patella has polished faces to reduce friction to a minimum and one of these faces is called Lenoir's facet. This was described in 1891 by Camille Lenoir, of whom almost nothing is known, but the eponym is often wrongly attributed to Adolph Lenoir, a professor of anatomy in Paris at about the same time.

The finer structure of the bones has interested later generations of medical men, for powerful hand lenses and then the microscope were required to make much sense of what, with the exception of the dental enamel, is the hardest of the body tissues.

By far the best-known name associated with the bones is that of Clopton Havers. His Haversian canals and Haversian systems have become an integral part of the vocabulary of osteologists. The Haversian systems consist of numerous cylinders of bone, each a fraction of an inch in diameter, laid down around a central channel, the actual Haversian canal, which carries blood and lymph vessels and nerves to supply the surrounding bone. From the central canals, tiny spidery channels come off at right angles. These are the canaliculi, or little canals, which communicate with the individual bone cells entombed in tiny cavities deep in the rigid substance of the calcified matrix of the bone.

The Haversian systems lie in rows parallel to the long axis of the bone shafts, and where the shape of the bone changes they lie at all angles and planes and are partly modified by the stresses which the bone is required to resist. The Haversian canals are themselves connected by other canals crossing to adjacent systems. Near the marrow cavity and at the outer surface they communicate with outside blood vessels and nerves. These cross channels, particularly those linking with the surface membrane, or periosteum, are known as Volkmann's canals. Havers and Volkmann described their canals at widely different points in time and place.

Clopton Havers was born in England in 1650. He studied medicine in the Netherlands and practised as a physician in London. His observations on bone were published in London in 1691, a decade before he died. Alfred Wilhelm Volkmann was born in Germany in 1800. He was professor of physiology in Dorpat and Halle in Germany, and described his canals in 1873.

The physiology of muscles, that part of the framework which gives the human body its mobility and bulk, is not yet fully understood. And since there is still scope for research into the contractility of muscle, it is hardly surprising that so complex a mechanism baffled earlier generations of investigators. The actual detailed description of the gross muscle mass was, however, mapped out long ago. The names of many muscles go back thousands of years. Galen described more than three hundred, noting and naming some for the first time.

As with other body structures, however, the first accurate detailed descriptions and illustrations of muscles were not to move forward until the great resurgence of descriptive anatomy of the Middle Ages. Leonardo da Vinci, who excelled in myology—the study of muscles—just as he did in osteology, made detailed studies of the muscles, both of their form and function. He discovered what various muscles did by pulling on their tendons, as Galen had done long before. He joined muscles by cords to demonstrate how they acted and he made models of lever systems to simulate the action of muscles on joints. He seemed to know about muscle tone, the partial tension that is maintained constantly during life, and realized that this was lost at death, when the muscles lengthen slightly. Leonardo also knew that muscles are usually arranged in antagonistic pairs, each working against the other, and that the antagonist relaxes to allow movement to occur.

The first good, overall description of muscle anatomy came, of course, with the publication of Vesalius's *De Fabrica* in the middle of the sixteenth century. It was the second volume which dealt with muscular structure and the engravings by van Kalkar are works of art as well as science. A series of living poses reveals the body in different stages of dissection, set against backgrounds of medieval Italy, with the town, ruins, rocks, and vegetation giving an immediacy in startling contrast to the functional and sterile pictures of modern textbooks.

In the Galenic tradition, Vesalius always tried to relate structure to function, and van Kalkar's drawings, especially those of the muscles, all reveal this dynamic approach. Other writers and their artists produced pictures which were sometimes more accurate in detail, but they were illustrations of dead, immobile cadavers, lacking the vitality with which Vesalius and van Kalkar imbued their figures.

Several of the illustrations show muscles which exist only in apes, such as the upward extension of the stomach muscles which can be seen when the abdominal wall is drawn in. In the higher apes, these muscles extend right up to the collar bones, but in the human they finish at the margin of the ribs. These rectus abdominis muscles have several transverse bands of fibrous tissue interrupting their length, which can be seen on any muscular person, as the rectus is

This schematic diagram, a fanciful and stylized view of muscle anatomy, appears in a manuscript, believed to date from the late thirteenth century, which deals with a range of medical topics.

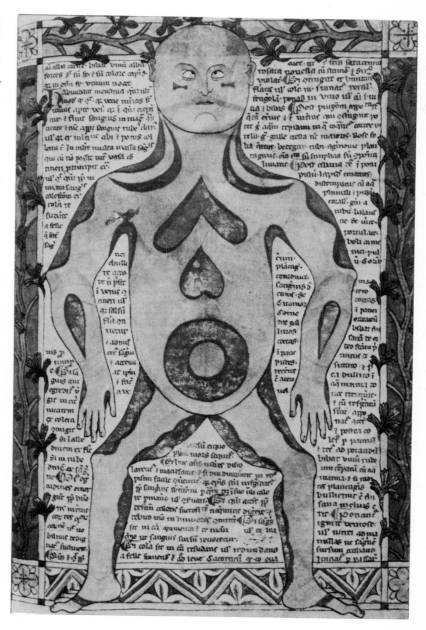

The second volume of *De Fabrica* is concerned with muscles. Having shown the superficial muscles in earlier plates, Vesalius used this and subsequent illustrations to reveal, layer by layer, the deep muscles of the body. Most of the figures are depicted against scenes of contemporary Italy.

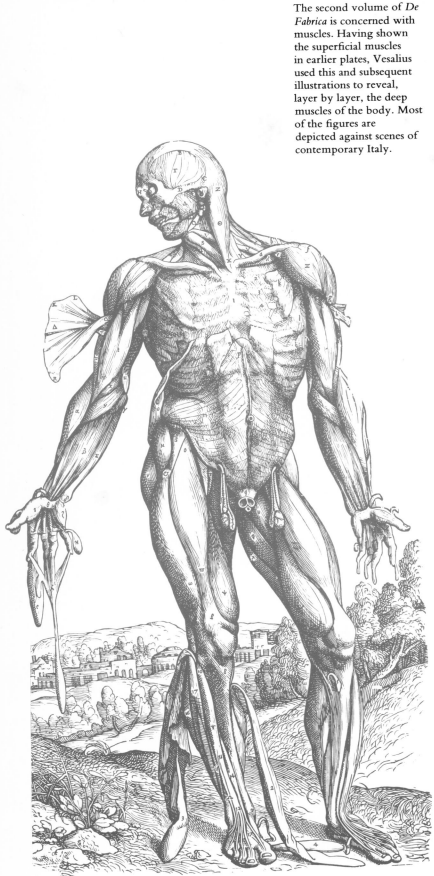

separated by them into three low humps. These are said to be remnants of the segmented origin of all vertebrates, a faint vestige of our evolutionary past. *De Fabrica* also shows an extra neck muscle which only apes possess, although Vesalius draws attention to this in his text. He made a number of errors, however, such as the description of a seventh muscle to move the eyeball, which does not exist in humans, and, strangely, he missed the muscle which raises the eyelid.

On the credit side, Vesalius not only made the best dissections and had the best illustrations of muscles up to his time, but he also conducted physiological experiments on muscle tissue. He showed that if a muscle is cut along its length, there is very little effect on its function; but a transverse cut puts a muscle out of action. He noticed, too, that a muscle cut across in this way retracts, because of its normally semi-contracted state. Vesalius also investigated the nervous control of muscles, showing that the contraction of muscles is stimulated through the nerves.

The seventeenth century saw further work on the physiology of muscles. Niels Stensen, the Danish anatomist who later became a Roman Catholic bishop, made a number of important observations. His was the first scientific work on muscles. Published in 1664, it gave details of the fibril structure as seen under the microscope. Stensen realized that the action and function of a muscle are the aggregate of all the contractions of its individual fibres. Richard Lower and Thomas Willis, the Oxford anatomists, went into print soon after this and confirmed Stensen's microscopic descriptions.

In 1677, the year of his death, Francis Glisson, the English physician who is best known for his work on the liver, showed that a muscle decreases in volume when it contracts, the opposite of what was previously thought. Glisson was a pioneer of the irritability theory of body function. His concept of irritability was to be used much later in connection with muscle, but in the sense of excitability. Whereas Glisson thought irritability was a property innate in all living tissue, almost a century after his death the Swiss physiologist Albrecht von Haller found that a muscle was "irritable" either spontaneously or when an external stimulus was applied. Mechanical or electrical stimulation of an isolated muscle—even soon after death—will lead to contraction, irrespective of its nerve supply. Heart muscle and the smooth muscle of the intestine have an intrinsic irritability; and single isolated heart muscle cells in tissue culture in the laboratory will contract rhythmically of their own accord.

In the eighteenth century, Giorgio Baglivi was the first to see the microscopic difference between skeletal muscle fibres, which are crossbanded or striated, and the plain, smooth fibres of involuntary muscle. A confirmed iatrophysicist, who compared the body to a series of machines, Baglivi proposed that striated muscle was designed for rapid, repeated activity and smooth muscle for slow, sustained contractions. In this he was fairly close to the truth.

Baglivi was a talented teacher in Rome and commanded large audiences of students. His teaching is epitomized by his statement: "Let the young know that they will never find a more interesting, more instructive book than the patient himself." He has been described as the master of Italian clinicians in his time and it was a tragedy that he died at thirty-nine.

Anton van Leeuwenhoek, the first systematic exponent of the microscope, gave the earliest detailed description of striped muscle fibre, which speaks volumes for his continuing good eyesight, for he was about eighty-three years old when he published this work in 1715. He described both the striations and the fine longitudinal fibrils, as well as the diaphanous sheath, the sarcolemma, around the fibre.

In 1738, Hermann Boerhaave, another Dutchman and one of the greatest names in eighteenth-century medicine, reissued in Latin a book by a compatriot which had never been noticed when it first appeared in Dutch in 1669. This was the *Bible of Nature* by the physician Jan Swammerdam which described instruments for measuring muscle function that are still used today.

In the nineteenth century, the major work on muscle was directed at understanding its chemistry. As early as 1800, there were suggestions that the splitting of chemical substances was responsible for muscle action. In 1845, the German physiologist Hermann von Helmholtz (a descendent of William Penn, the founder of Pennsylvania) worked on the chemical changes in frog muscle and decided that the major source of body heat comes from these reactions in the muscles. He was the first to use a thermocouple to measure temperature changes in isolated muscles.

It was already known that there are electrical currents in contracting muscle and by 1848 it became known that active muscle becomes acid. This observation was confirmed by von Helmholtz in 1850, the same year that lactic acid was first synthesized. This was the substance identified at the end of the century as the acid produced in muscle, a finding contributed by the French physiologist Claude Bernard.

It has been a long climb from the struggles of the ancients to comprehend animal movement to the complex electron-microscope ultra-structure understood today. And the pursuit of more facts about the complex chemistry of muscle action still continues.

Over the centuries, muscles, like bones, have attracted relatively few eponyms, undoubtedly because most were given Latin or Greek names

Niels Stensen, an important member of the European scientific community of the seventeenth century, was born in Copenhagen in 1638. He studied medicine under Bartholin in Denmark and under Sylvius at Leyden before beginning his work as an anatomist in Amsterdam. In 1669, he became a Roman Catholic and Pope Innocent XII appointed him to a titular bishopric. As well as being a distinguished anatomist who made significant discoveries about muscle and brain function, Stensen was also a crystallographer and is known as the Father of Geology. He died in Germany in 1686.

Hermann Boerhaave preached the Hippocratic doctrine that the patient is the centre of the doctor's universe. Born in 1668, Boerhaave became professor of medicine and of botany at Leyden. A superb teacher, he attracted students from all over Europe to his twelve-bed clinic. He had an ornately bound book which he claimed contained all the secrets of medicine. After his death, in 1738, his colleagues and pupils opened it with great anticipation. It was found to contain blank verse—and the dictum "Keep the head cool, the feet warm, and the bowels open."

Albrecht von Haller, one of the most gifted men of science in the eighteenth century, was born in Berne in 1708 of an aristocratic family. He began his university studies at Tübingen at the age of fifteen. From there he went to Leyden where he studied with Boerhaave. In 1735, he was appointed professor of anatomy, surgery, and botany at the newly founded University of Göttingen. Von Haller dissected more than four hundred cadavers and carried out research on the function of muscles and nerves. He compiled an eight-volume edition of all the major physiological discoveries to that time. He died in Berne in 1777.

which related to their functions. Nevertheless, there are several examples of conflicting claims. No less than three different men are involved with the muscle in the mouth which compresses and purses the lips. Karl Krause (father of the much more eponymized Wilhelm) wrote about this muscle in 1842, and it became known as the sucking muscle of Krause; Hungarian-born Edward Klein described it twenty-seven years later, when it duly became Klein's muscle; and the same muscle reappeared in the guise of Aelby's muscle following its description by Christopher Aelby in 1878.

Only two men are involved with the lateral scalene, which is a small and inconstant muscle running between the spine and the first rib, at the root of the neck. Originally named for Bernhard Albinus, who gave the first description

Leonardo made a thorough study of the structure and function of muscles and many of his explicit drawings show the action of muscles. Particularly clear and accurate are his illustrations of the shoulder muscles.

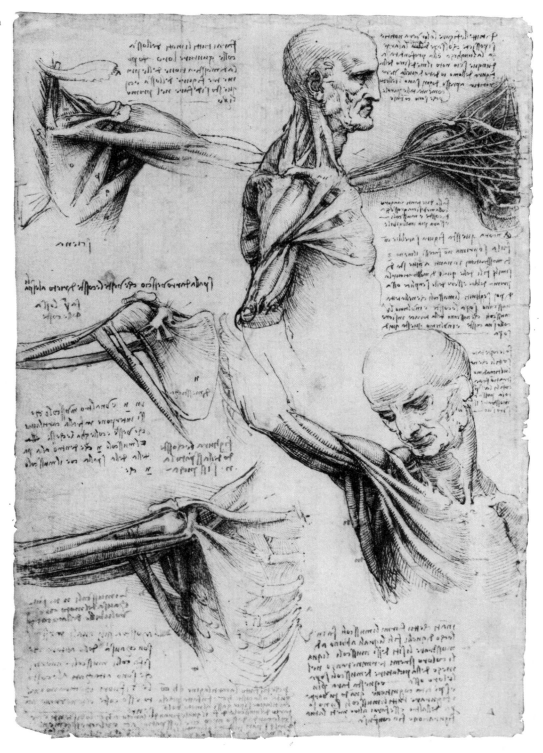

in 1734, it later became known as Gibson's muscle after Francis Gibson, an English professor of medicine who described it again in 1846.

The name of Bernhard Albinus is, however, attached to other muscles, including the rhomboid—diamond-shaped—muscle at the side of the nose, which he described in 1747. This is fitting because Albinus, who was born in Germany in 1697, was professor of anatomy and surgery at the University of Leyden for fifty years. He edited the works of Harvey, Vesalius, Fabricius, and Eustachius, in addition to his own numerous publications which were noted for their style and for the beauty and accuracy of the illustrations executed by the artist Jan Wandelaer. Albinus was the first to classify and arrange the muscles in a proper manner and many of the names he gave to them are still used.

The large tendon at the back of each ankle is sometimes called the hamstring because in a butchered animal it forms the sinewy part of the fleshy ham muscles. The hamstring tendon in the heel is known as the Achilles tendon and is one of the most well-known eponymous structures in the human body. According to legend, Achilles, the Greek hero, was made invulnerable when his mother held him by the heel and dipped him into the River Styx. But since the water did not touch the heel by which she held him, this area was not rendered invulnerable and it was at this spot that Paris succeeded in inflicting a mortal wound.

A little-known alternative name for the Achilles tendon is the tendon of Hippocrates. Once again, as in the case of Vesalius and Fabricius, it seems unjust that the name of the Father of Medicine should linger only in one rather insignificant and little-used eponym.

The small area between the legs seems to have generated a disproportionate amount of myological eponyms. This perineal area has a whole series of named muscles, mostly relating to the anus and the male genito-urinary apparatus. The names commemorated in this intimate region include those of Wilson, Theile, Treidig, Sebileau, Rudinger, Roux, Lesshaft, Kolrausch, Houston, Kobelt, Hoffmann, Ellis, and Braune. It seems that where organs of special sense or particular function are concerned anatomists rush in with far more enthusiasm than is generated by muscles of the limbs or of the trunk. This attention may well be related, however, to the more practical concerns of surgical anatomy.

Many muscles have retained the lengthy Latin names which were attached to almost all body structures until the revolution in anatomical nomenclature. A well-known anecdote involves one of these muscles, now known to anatomists simply as the grace muscle. The story concerns a professor of anatomy, who was invited by a colleague in another speciality to dine in the hall of an Oxford college. On arriving at table at this rather formal occasion, the anatomist was suddenly discomfited by a request to say grace to the assembled fellows. Neither a religious man nor a Classics scholar, he had no idea of any appropriate words. His wit was as quick as his Latin was poor, however, and, folding his hands reverently before him, he solemnly chanted: "Musculus levator labii superioris alaeque nasi." The other diners were suitably impressed by this novel incantation—a rendering of the Latin name of the muscle which raises the upper lip and dilates the nostrils when snarling.

Fibrous tissue is sometimes ignored as an integral part of the framework, but if the human body could be viewed through some magic lens that rendered invisible everything but this tissue, the result would be spectacular. The whole body, with the exception of much of the nervous system, would be seen in a translucent, diaphanous form. For it is this fibrous tissue which sheathes virtually everything, from the layer under the skin to the compartments around every muscle, from the membranes around the organs to the thin cladding on each bone. Without the support of fibrous tissue, the human body would slither to a heap of jelly hanging around the ankles, the muscles drooping off the bones, the organs shapeless masses.

Where the sheets of fibrous tissue are extensive, they are given the name fascia, from the Latin word for a bundle, because the tissue consists of myriads of microscopic fibres of a tough protein called collagen, often mixed with some stretchy elastic fibres. It is this collagen that makes meat tough; the older the animal, the more fibrous tissue tends to accumulate between the muscle bundles that we call meat.

This ubiquitous material, which keeps the human body from falling apart, has been a glorious haven for anatomical name-droppers. Anatomists and surgeons by the score are commemorated in sheets of fascia or various bands, ligaments, holes, arches, spaces, and triangles which are left here and there about the body in the complex architecture of the fibrous network. They include Alcock's canal, Bryer's bursa, Zang's space, the foramen of Winslow, Sharpey's fibres, Imlach's ring, and Hunter's canal.

Among the most prominent fascial structures, however, are the arches of Haller, curved doorways in the diaphragm, which is the large fibromuscular sheet separating the chest from the abdominal cavity. The arches of Haller are found at the back of this domed sheet, which has a number of openings for various tubes and vessels. The eponym, which recalls the eighteenth-century Swiss physiologist, anatomist, and botanist Albrecht von Haller, is nonetheless a poor monument to so illustrious a figure.

Anatomy actually begins on the outside of the body with its visible form. Surface anatomy is important not only to surgeons, who must know where to cut, but also to artists who seek to capture and reproduce the infinite detail of the human body. Indeed, some of the credit for the revival of interest in anatomy in the fifteenth century is due to painters and sculptors, especially to the great masters of Italian art. Far earlier than this, however, there is evidence of artistic appreciation of muscle anatomy. From Knossos in Crete, a fragmented Minoan statue, dating from about 1500 B.C., has a forearm in which the modelling under the surface reveals at least seven different muscles, all exactly in the correct position. The sculptor obviously studied a real arm in every detail.

The ancient Greeks, too, must have had a great knowledge of surface anatomy to be able to sculpt their masterpieces. Although their expertise could not have been gained from dissection, which did not take place so early as the fifth century B.C., their appreciation of surface musculature was profound.

A statue from Argive Hereum, for example, dated to about 450 B.C., clearly shows the outline of the pectineus muscle on the upper part of the thigh. It was always believed that knowledge of this particular muscle could only be obtained from dissection, but recently it has been demonstrated that, in certain postures and with the right lighting, the pectineus can actually be seen in a living person.

Surface anatomy was largely ignored throughout the Dark Ages and until the late Middle Ages. In the fifteenth century, when painters and sculptors turned from the flat, lifeless art of medieval times, in their effort to achieve dynamic realism of form and posture, they began to take intense interest in the human body. In Andrea Mantegna's painting *Dead Christ,* for example, the experimental foreshortening of the body as viewed from the feet involved a wealth of knowledge of surface anatomy.

It is only in the recent past, and especially with the discovery and publication of Leonardo da Vinci's notebooks, that the superlative anatomical achievements of the Renaissance artists have come to light. Nevertheless, surface anatomy must have the longest pedigree of any subspeciality of anatomy, for the exterior of the body is accessible to anyone who cares to study it. Certainly it had been a preoccupation with the surgeons seeking lines and landmarks to assist them in the perfection of their technique.

Most well known of the men commemorated in surface anatomy is the nineteenth-century Viennese anatomist Karl Ritter von Edenburg Langer, whose Langer's lines are familiar to every student of surgery. Yet what is named after Langer is virtually invisible, for the eponym refers to the tension, or cleavage, lines which occur within the skin over the whole body surface. There is a complete network of fibres immediately underlying the skin, and these run in bands and bundles according to the mechanical stresses and angulations at various points. These fibres give strength and elasticity to the skin and help it to accommodate to the changing shape of the body in different postures.

Langer's lines are of great practical importance to the surgeon, who always tries to cut along rather than across them. The creases in the skin, for example, run like a collar, rather than vertically up and down. When the head is thrown back, these creases pile up at the back of the neck, but when the chin is tucked down, the surface layers pucker in front. A surgical operation on the neck must be done through an incision parallel to Langer's lines, so that the resulting scar will be as inconspicuous as possible. This is an important consideration in such a delicate area as the front of the neck. If the incision were made across those lines, the healing scar would pucker at every nod of the head. It might even split open and would almost certainly heal with a wider, more ugly scar.

Langer's lines have been mapped over the entire skin area. To achieve this, Langer made holes in the skin of dissecting-room corpses and observed the direction in which the skin split. This told him the way in which local tension lines were directed. By painstakingly repeating this process over every part of the body, he was able to chart stress lines for the entire surface.

In 1500 B.C., Minoan craftsmen had knowledge and appreciation of muscle anatomy, as is evident in this detail, of a man tethering a bull, from a solid gold cup which was found in a tomb in Greece.

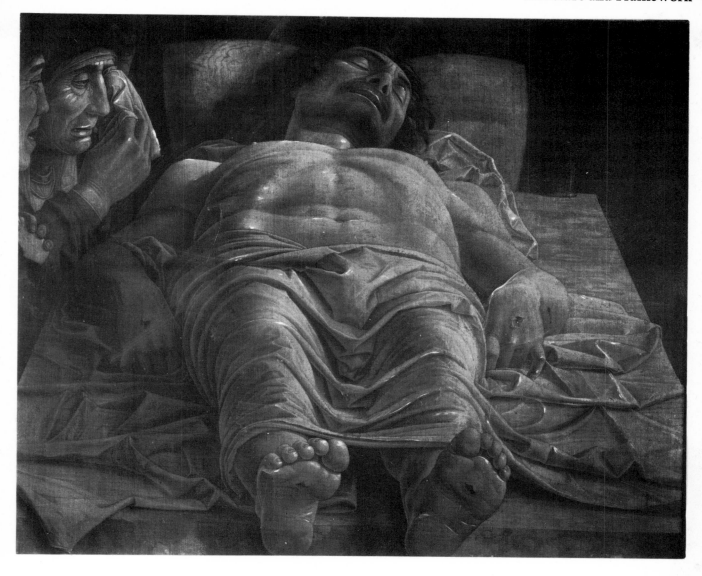

This important work was published in 1862.

A more specific line—the furrow in the crease of the groin, seen when the thigh is flexed at the hip—is known as Holden's line. This is a surgical landmark for the many operations in the groin area, where hernias are particularly frequent. It was described by Luther Holden, a surgeon at London's St. Bartholomew's Hospital in 1877, two years before he became president of the Royal College of Surgeons.

Many other lines on the body surface are actually invisible lines between specific points which serve as guides for surgeons. They are also of help to anthropologists who make comparative studies of skull sizes and proportions. Reid's base line, for example, is the horizontal line which runs from the ear canal to the lower edge of the eye socket. A reference marker for other dimensions of the skull, it was devised in 1884 by the Scottish anatomist Robert Reid.

Bryant's triangle helps in the diagnosis of fractures of the neck of the thigh bone and was described by Sir Thomas Bryant, the nineteenth-century English surgeon. Petit's triangle, on the flank, was described by the French surgeon Jean Louis Petit as "the triangle of lumbar hernia" in 1705. The famous Scarpa's triangle, important for surgery of the groin, was described in 1817 by Antonio Scarpa, the Italian anatomist.

Morris's kidney box is a surface marking made on the skin to indicate to the renal surgeon where the margins of the kidney lie beneath. This was described in 1893 by Sir Henry Morris, who was surgeon to the Middlesex Hospital in London and the president of the Royal College of Surgeons between 1906 and 1909.

Charted many centuries after the visible form of the human body was immortalized by the Renaissance artists, the invisible lines of surgical anatomy are as important in therapeutics as is a sound knowledge of the underlying tissues. And it was with the structural tissues—the physical framework—that the whole science of descriptive anatomy began.

In the fifteenth century, the revival of naturalism in art brought an intense awareness of physical structure and form. Notable for its anatomical perspective, Andrea Mantegna's painting of the dead Christ was almost certainly the model for a much later painting by Rembrandt that used the same perspective.

The Heart and Circulatory System

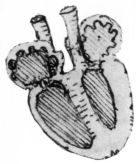

The earliest reasonably accurate drawings of the heart were those of Leonardo da Vinci in the fifteenth century. His only mistake, as in the drawing above, was in reproducing the mythical pores in the septum which had been postulated by Galen.

Of all the body systems, the heart and circulation presented the most difficult of anatomical puzzles to investigators of the past. The fantastic theorizing about the circulatory system, begun two and a half millennia ago, led to an almost complacent acceptance of Galen's nonsensical explanation until the seventeenth century.

Far from Greece, the ancient mainstream of anatomical inquiry, the Chinese alone of the old civilizations have a tenuous claim to priority in the discovery of the circulation of the blood. The

most interesting statement in recorded ancient Chinese anatomy is the mention in the *Nei Ching* (believed to have been a digest of earlier works compiled in about the third century B.C.) that "the heart regulates all the blood in the body . . . the current flows continuously in a circle and never stops." Although this comes close to the truth ultimately revealed in the West in the seventeenth century, it was probably no more than an inspired guess, since dissection was virtually unknown in ancient China.

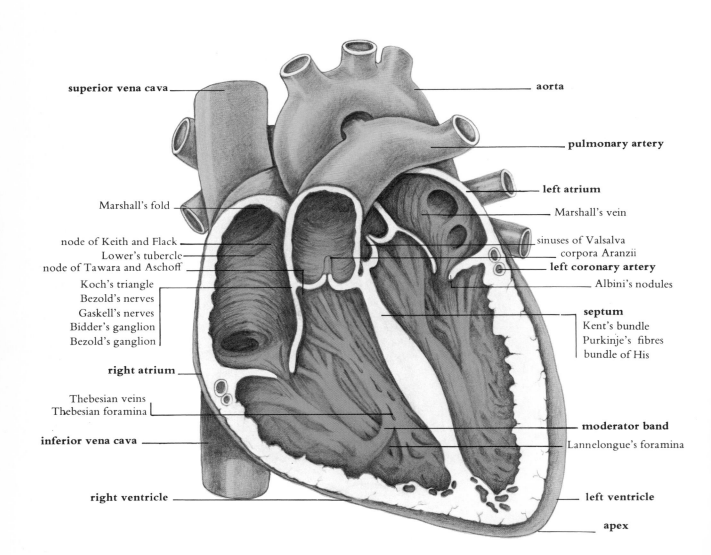

The Chinese were, however, great exponents of pulse-taking as a diagnostic aid. Doctors often spent several hours simply taking the pulse, of which they recognized more than two hundred variations.

The ancient Greeks were the first to consider seriously the nature of the heart and blood vessels. From about the fifth century B.C., theories about the blood and its movements became inextricably linked with the concept of the four elements and the pneuma, which were thought to pervade the whole universe and which, in addition to the four bodily humours, were believed to be the essence of life.

Although there are hints of the humoral theory in ancient Chinese, Hindu, and Babylonian writings, the concept was put firmly on the road by the Greek philosopher Empedocles who lived in Sicily in about 480 B.C. He maintained that the entire universe was constituted solely of the four elements—fire, earth, air, and

water. Pneuma, the life force, pervaded all matter. In the human body the seat of the pneuma was the heart from where it was distributed to all the tissues. Empedocles believed that the body breathed in pneuma from the air, not only through the lungs, but through tiny pores in the skin.

The later, Hippocratic school reconfirmed and elaborated upon the humoral theory, which was to hold sway for the next two thousand years. Health and sickness were due to a balance or imbalance of the four humoral substances in the body—blood, phlegm, and black and yellow bile. The essential factor for life was the innate heat which resided in the blood and was fuelled by food and drink in combination with the ubiquitous pneuma. Hippocrates conveyed the idea of the circulation in the midst of all these hypothetical substances when he wrote, "The vessels which spread themselves over the whole body, filling it with spirit, juice, and motion,

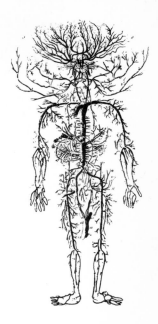

Although Harvey published his work on the circulation of the blood in 1628, by the eighteenth century the network had still not been mapped in detail. This engraving of the arterial system, from a French encyclopaedia of 1745, appears bizarre when compared to a modern diagram.

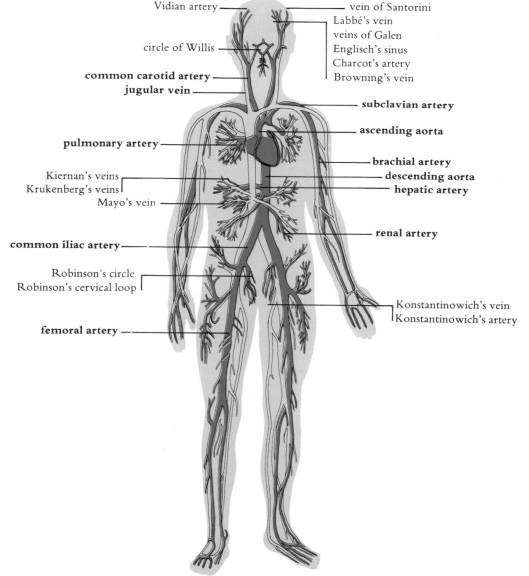

Vidian artery

circle of Willis

common carotid artery
jugular vein

pulmonary artery

Kiernan's veins
Krukenberg's veins
Mayo's vein

common iliac artery

Robinson's circle
Robinson's cervical loop

femoral artery

vein of Santorini
Labbé's vein
veins of Galen
Englisch's sinus
Charcot's artery
Browning's vein

subclavian artery

ascending aorta

brachial artery
descending aorta
hepatic artery

renal artery

Konstantinowich's vein
Konstantinowich's artery

35

are all of them but branches of an original vessel. I protest, I do not know where it begins or where it ends, for in a circle there is neither beginning nor end."

The arteries were said to contain air. Originally the Greeks had used the word artery only when referring to the windpipe or other air passages, but the word later came into use for what are now known as arteries simply because at examination after death these vessels were observed to be empty. The veins, however, were engorged with blood, and, not unnaturally, they were presumed to be the only canals for blood.

It was Aristotle who placed the heart firmly at the centre of all vital processes. Moreover, he insisted that it was the seat of intelligence. He made no distinction between arteries and veins and described the heart as three-chambered, an error undoubtedly made because he dissected only lower animals and assumed their three-chambered heart to be universal. Although Aristotle was the first comparative biologist, and one of the best, he further consolidated the humoral theory, adding the four essences, or qualities, of hotness, coldness, dryness, and wetness and integrating them with the elements and humours to make an even more complicated and dynamic theoretical system. In some ways, if his theory is considered in modern terms as the complex of endocrine agents, biofeedback mechanisms and other components of the *milieu intérieur*, then Aristotle had come remarkably close to the truth.

The next advances in the understanding of the heart and circulation were made in Alexandria by that rival pair Herophilus and Erasistratus. Herophilus restored the intelligence to the brain where it belonged and he made a clear distinction between the thicker-walled arteries and the veins. He studied the pulse and recognized it as an active process, but mistakenly assumed that it was the arterial wall itself that contracted as the result of some inherent mechanism, rather than a pressure wave from the heart causing the vessel to expand passively.

Erasistratus considerably developed the pneuma theory, especially in connection with the heart. He maintained that pneuma was taken into the lungs and passed down the vessels to the heart, where it became changed into vital spirit that was then distributed throughout the body by means of the arterial system. The blood, according to Erasistratus, was a separate system, ebbing and flowing in the veins. When the vital spirit reached the brain, it was converted in the cerebral ventricles, the spaces in each hemisphere, into another kind of pneuma, animal spirit, which was then conveyed to all the tissues through the hollow nerves.

Almost two thousand years before microscopy revealed the existence of a capillary system, Erasistratus came amazingly close to

Representations of the heart, like the four above, were often used on Egyptian amulets and passed into the Egyptian system of hieroglyphics.

discerning the truth about the capillaries, the tiny blood channels which permeate the tissues and link the arterial system with the veins. He claimed that all arteries, veins, and nerves subdivided endlessly until they were too small to be seen and that the final minute endings actually formed the flesh of the body.

His explanation why arteries bleed copiously during life, but are empty after death was ingenious. He said that when an artery was cut during life, pneuma immediately escaped, thus leaving a vacuum. This was quickly filled by blood which was sucked from the veins through the small terminal tubes. For this to happen, there had to be terminal connections between the arteries and veins, which is exactly what capillary beds consist of.

Dissection was practised extensively in Alexandria and, consequently, advances were made in understanding the anatomy of the heart. Erasistratus decided that the right ventricle of the heart filled with blood during the relaxed phase of the heartbeat and that the left side filled with pneuma. Both blood and pneuma were expelled during the heartbeat and their return was prevented by the semilunar valves at the base of the aorta and the pulmonary artery. He was more accurate than later investigators who thought that the heart sucked like a syringe during relaxation, instead of blowing during contraction.

Erasistratus named the tricuspid valve between the right atrium and the ventricle and correctly divined its function. But he believed that the bicuspid, or mitral, valve between the left atrium and ventricle ensured that the pneuma, or by then vital spirit, only left the heart by way of the aorta, instead of returning to the lungs. Thus he came close to the concept of a one-way, and necessarily, a circulating, blood system.

Many parts of the circulatory system were described by Erasistratus, including all the major vessels emerging from the heart, and the intercostal, renal, gastric, and hepatic arteries. He also described many of the major veins, including the azygos vein in the chest.

In about A.D. 100, Rufus of Ephesus, the Roman physician and a prodigious writer, made the observation that the apex beat (when the heart strikes the inside of the chest wall) is felt during contraction, or systole, and is simultaneous with the arterial pulse. This important discovery was almost completely overlooked for centuries afterwards. It was assumed that the pulse occurs when the heart is relaxed.

Little progress was made towards understanding the heart and circulation until the second century A.D., when Claudius Galen expanded earlier theories into a durable, if tortuous, hypothesis. Ingeniously, he managed to enmesh all the various past suppositions into one complex system, adding a few embellishments of his own.

According to Galen, the starting point of the circulatory system was the gut, from where he believed absorbed food in the form of liquid chyle was carried by the portal vein to the liver. (Chyle was Galen's word for the milky emulsion and it is still used.) Galen considered the liver the focal organ of the system, more important than the heart, for it was in the liver that chyle was converted into blood. This process was brought about, he claimed, by the addition to the chyle of natural spirit, yet another variety of pneuma. From the liver the blood travelled through the veins, taking nutrients in the form of this natural spirit to all the tissues. When the blood was depleted of the spirit, it then flowed back along the same channels to the liver. And so the cycle of replenishment was repeated.

Galen realized that this hypothesis satisfied the question of sustenance for the tissues, but that it did not account for the functions of intellect and nervous activity. To work this aspect of the life process into his grand scheme, he theorized that blood entering the right side of the heart through the vena cava gave up the impurities it had acquired from the tissues as it ebbed and flowed in the veins. Here, in the right ventricle, the way in which these impurities were voided was by mixing with air, which came from the lungs down the "artery-like vein," which is now known as the pulmonary vein. The impurities were then exhaled with the outgoing breath and the purified blood then ebbed back to the liver to begin a new cycle. A smaller proportion of the blood which went up to the right ventricle, however, trickled through small "pores" in the thick septum, or wall, which separates the right and left ventricles. That no such pores actually exist was no deterrent to Galen who apparently decided that for this theory to work they had to exist and therefore they did exist.

The blood that filtered through these mythical pores entered the left ventricle and came in contact with pneuma, brought in from the outside air through the windpipe and lungs, and reached the left side of the heart by way of the "vein-like artery," which is now known as the pulmonary artery. This contact between pneuma and blood turned into an active medium called vital spirit, which was distributed throughout the body by the aorta and other arteries. Galen explained that whereas the venous blood with its rather low-grade natural spirit was dark, thick, and sluggish, the high-quality, vital-spirit-charged blood from the left side of the heart was bright red, active, and spurting, fit for the nutrition of the brain. After the vital spirit arrived at the base of the brain, it was distributed, according to Galen, by the rete mirabile, a meshwork of blood vessels inside the skull, present in some animals, but another merely mythical structure in man.

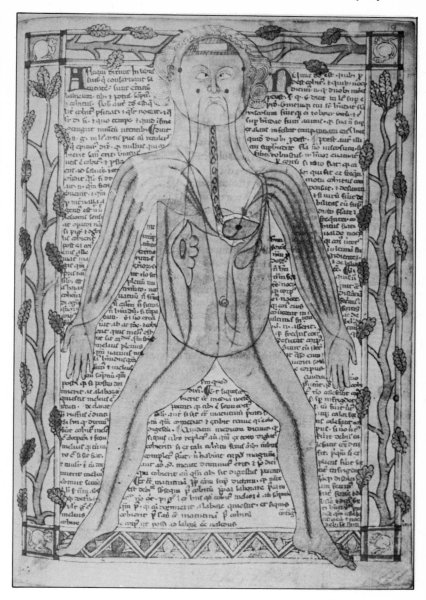

On surer ground, however, Galen did establish for the first time that arteries contain blood, not air. To demonstrate this, he tied off a length of animal artery with two spaced ligatures. When the vessel was cut between these two ligatures, only blood escaped. He concluded that, since this blood could not have entered the artery to replace lost pneuma, as Erasistratus had suggested, then it must have been present in the artery all the time.

But Galen's absurd humoral hypothesis was to endure until the time of William Harvey and beyond. It has been said that Galen might just as well have discovered the circulation of the blood as Harvey, but this disregards the absence in Roman times of the range of technology available to assist people's understanding of complex body systems. Galen actually used familiar models where they existed. The mixing of chyle with blood in the liver, for example, suggests a

In the early medieval period, the arteries—defined as "veins that pulsate"—were said to arise from a black grain in the heart, as shown in this illustration from a thirteenth-century medical manuscript. The point of origin of the arteries was believed to bring them in close contact with the *spiritus*, also said to dwell in the heart, which was thought to connect directly with the trachea.

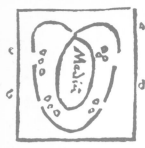

In an attempt to reconcile the views of Aristotle and Galen, Mondino de Luzzi described the heart as having three ventricles. This error, contained in his *Anathomia* of 1316, was still being perpetuated in editions of the book published two centuries later. The diagram above, from a 1513 edition, shows the three chambers, with the heart valves crudely drawn in on each of the two outside ventricles. In contrast, the rendering of the heart below, from a manuscript by Jacob Berengar of Carpi, published in 1523, is a model of sophistication. It shows the left ventricle open to reveal the aorta. The spiral structure at the top of the drawing represents the left auricle.

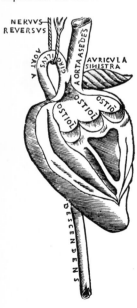

process such as cooking or brewing. The heat-producing, active process in the heart is reminiscent of smelting. But there were no mechanical models, such as the pump—a popular modern analogy for the heart—to put even a hint of a pressurized circulating system into his mind.

Although the functioning of the cardio-vascular system was totally uncomprehended, much of its anatomy was named by the third century A.D. The men who described many of the structures are unknown, but the Latin and Greek names they bequeathed to them were derived from objects familiar in everyday life. Some Latinized terms were, however, later coined during the heyday of the Renaissance, when Latin was the universal language of academics. The name ventricle, for example, used for the major paired chambers of the heart, means belly, just as the frontal aspect of the body, where the actual belly lies, is called the ventral surface.

The atria, the small upper chambers of the heart, recall the atrium, the entrance chamber of a Roman house; here the analogy is that the blood waits in the atrium before being admitted to the ventricles. At each side of the atria there is a projecting pouch called the auricle, because of its imagined resemblance to the ear. The word valve refers to the leaf of a folding door—*valva*, in Latin—and the resemblance to the mobile flaps is obvious. The free edges of the movable flaps of these valves are called cusps, meaning pointed ends. Behind the valve flaps of the semilunar valves are slight cavities called sinuses, from the Latin for a recess.

Until the Renaissance, when more of the anatomy was seen and comprehended, there was not much increase in understanding. Even during the early Middle Ages, medical men, still overshadowed by the towering figure of Galen, could add little to the corpus of knowledge. Indeed, sometimes they seemed to go backwards. The Italian anatomist Mondino de Luzzi, for example, used a diagram of the heart in his textbook of 1316 which showed two ventricles on the outside and one in between.

Leonardo da Vinci was the first to produce accurate illustrations of the heart, and some of his drawings are masterpieces both of art and medical instruction. It was he who recorded the moderator band, the strip of muscle which crosses the cavity of the right ventricle, providing support and carrying a branch of the conducting system, which was not even guessed at by others in his time. He made working models of the heart valves showing how they allowed blood to pass in only one direction. It would seem that this observation alone should have killed off the Galenic notion of ebb and flow and should have suggested ideas of the circulation to the mind of the genius Leonardo. But clearly

the time was not yet ripe for such heretical innovation, and even Leonardo drew in those non-existent pores in the septum because Galen had said they were there. Much of Leonardo's brilliant work was lost until recent times, however, and so his important observations had little influence on the evolution of anatomy and physiology.

As the result of anatomical dissection, rather than speculation, in the sixteenth century, anatomists like Dryander of Marburg finally described the heart and blood vessels more or less as they are. Yet they still had no idea how the system really works.

Charles Estienne, the sixteenth-century French physician, was the first to remark on the presence of valves in the veins—crucial to the later understanding of the circulation—but he had no notion of the role of the valves in preventing back-flow of blood and his concept of the vascular system was actually sketchy. Alessandro Benedetti, a professor at Padua, first applied the word valves to the structures in the heart, although they had been noted and described long before.

Even Vesalius contributed less to the understanding of the heart than to the other bodily systems. He described, and provided excellent illustrations of, the heart and blood vessels, but gave no further insight into their function, except to contradict Galen rather hesitantly. In the second edition of *De Fabrica*, published in 1555, he mentioned that he tried to put bristles through the pores in the septum but failed.

When Vesalius retired from his professorship at Padua, his pupil and assistant Matteus Realdus Columbus took over. He made an important discovery about the circulation. This was that the blood flows from the lung to the heart through the pulmonary vein. Actually, exactly the same discovery about the lesser, or pulmonary, circulation, had been made by Michael Servetus in 1553, six years before Columbus published. But Servetus perished at the stake and his book was burned, too, in the same year Columbus's book was published. It is not certain whether Columbus knew of Servetus's ideas when he formulated his own account, but in any event he was the first to demonstrate it experimentally.

Columbus also observed, during animal vivisection, that contraction of the ventricles of the heart is synchronous with the pulse wave in the arteries. This apparently simple observation reversed the long-held opinions of the ancients, with the exception of Erasistratus, who had said that the heart sucked blood in when it relaxed, instead of actively forcing blood out when it contracted.

A later occupant of the famous chair at Padua was Fabricius. He, too, made a small but fundamentally important contribution, in a tract entitled *On the Valves in the Veins*. Although a number of earlier anatomists had mentioned these little valves, Fabricius was the first to draw specific pictures of them. Yet Fabricius had no real idea of their function; he thought that perhaps they slowed the blood down as it flows outwards, to avoid it pooling at the extremities.

During the last few decades of the Galenic stranglehold on scientific thought, one man was at work whom Italians still regard as the discoverer, before Harvey, of the circulation. Andrea Cesalpino was a professor of medicine at Pisa during the second half of the sixteenth century and, indeed, some of his writings have suggested that he had stumbled upon the truth about the circulation of the blood. As early as

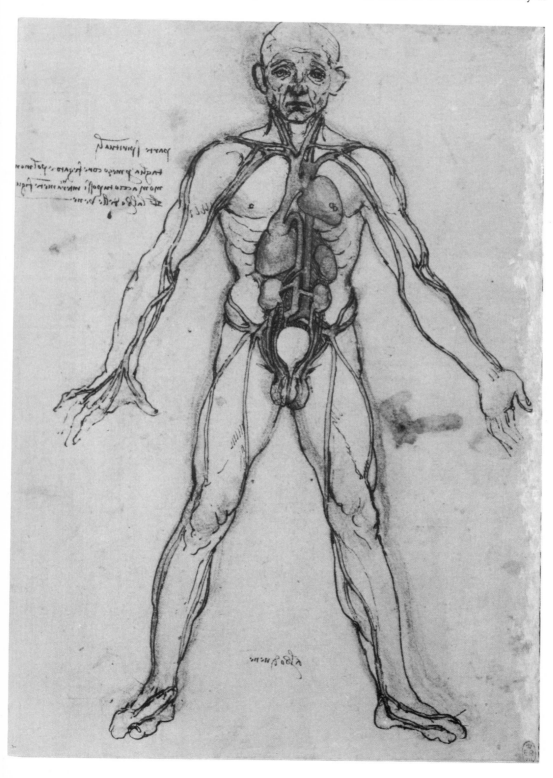

Leonardo took particular interest in the heart and circulation. He conducted his own experiments on the movements of the heart, and built special models to demonstrate the action of the heart valves. Besides detailed drawings of the blood supply to the limbs, his notebooks also contain separate illustrations of the arterial system in the male and the female.

1559 he had used the word "circulation." He contradicted Galen in a number of ways, not least by asserting that there was only one vital spirit, not three. But he wrote: "The orifices of the heart are made by nature in such a way that the blood enters the right ventricle by the vena cava, from which the exit from the heart opens into the lungs. From the lungs there is another entrance into the left ventricle from which, in its turn, opens the orifice of the aorta. Certain membranes are placed at the openings of the vessels to prevent the blood from returning, so

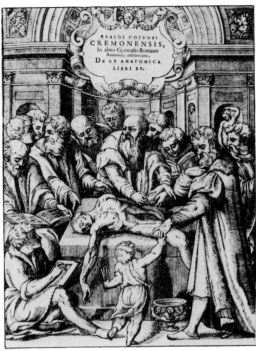

In *De Re Anatomica*, published in 1559, Realdus Columbus described the function of the lungs and provided Harvey with a key to the circulation.

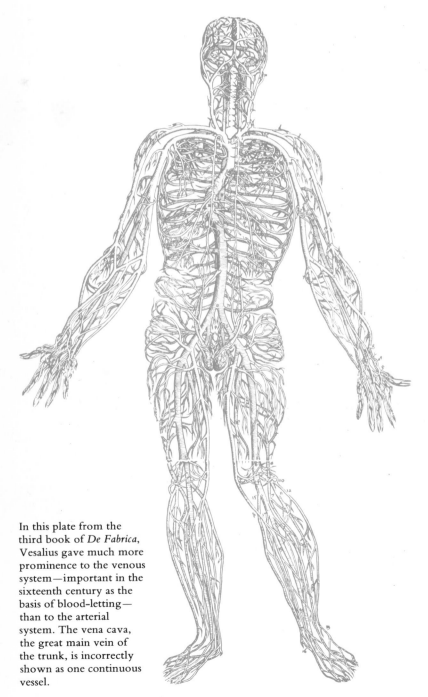

In this plate from the third book of *De Fabrica*, Vesalius gave much more prominence to the venous system—important in the sixteenth century as the basis of blood-letting—than to the arterial system. The vena cava, the great main vein of the trunk, is incorrectly shown as one continuous vessel.

that the movement is constant from the vena cava through the heart and through the lungs to the aorta." After all this good, factual description, Cesalpino pretty well lost his claim to priority by stating in his last book, published in 1606, three years after his death, that "the blood goes forth from the heart not only through the aorta and pulmonary artery, but also through the vena cava and pulmonary veins." This showed that he was still basically an ebb-and-flow man of the Galenic school.

No one could accuse William Harvey of rushing into print with his momentous conclusions about the circulation of the blood. He announced it to the scientific world at a lecture in London in 1616, but it was another twelve years before he was ready to publish his findings. This conservative approach was typical of Harvey, who cautiously worked over every aspect of a problem and proved each stage by experiment.

Although his notes are available, little is known about the timing or the sequence of the various stages in his dogged search for the great cardio-vascular truth. No doubt his four years under the tutelage of Fabricius at Padua set him on the right road, but the whole Harvey saga was one of contemplation of other men's work, the addition of his own labours, and an outstanding ability to dovetail all the available facts into the inexorable conclusion that in those days was as daring as it was momentous.

Indeed, Harvey himself admitted in the 1616 Lumleian Lecture at the Royal College of Physicians, that he feared the consequences of his declaration that the blood circulated. He declared, "But what remains to be said upon the quantity and source of the blood which passes, is of character so novel and unheard of, that I not only fear injury to myself from the envy of a few, but I tremble lest I have mankind at large for my enemies, so much doth wont and custom become a second nature. Doctrine once sown strikes deeply its root and respect for authority influences all men. Still, the die is cast and my trust is in my love of truth and in the candour of cultivated minds."

Many people had come close to anticipating Harvey's discovery. Hippocrates, hundreds of years before, skirted the matter with breathtaking closeness. Vesalius in the preceeding century had noted the valves in the veins and even observed that the great valves in the heart ensured that the blood could travel in one direction only, but never made that next great step of wondering where the blood went, if it could never return. Columbus, Leonardo, Servetus, and Cesalpino all came near the truth, too, but each backed away for want of that extra vision possessed by Harvey.

The Englishman's strength lay in his ability to concentrate on one problem at a time, yet keep the major mystery ever in sight. He refused to be sidetracked by all the old philosophical arguments about vital spirits and humours and pneuma and innate heat and the nature of life. He stuck essentially to the question of why there were valves in the vessels and whether their sole purpose was to create a one-way system.

Harvey, although not for a moment overtly denying Galen and the sacrosanct words of the ancients, kept his eyes on the problem. Unlike Vesalius, who in a brief supernova of brilliance finished his life's work in four or five years before he was thirty, Harvey worked for years, patiently putting together a cast-iron case for the circulation of the blood, testing and retesting it before allowing it to burst like a bombshell on the medical world.

The sequence of steps in Harvey's momentous discovery began with his assumption that the heart's active phase was during contraction, not relaxation. This had been well known to others, but Harvey was the first to prove it. He showed that a cut artery in a live animal spurts blood synchronously with the heartbeat. In contraction, the heart becomes harder, like a limb muscle flexing, and this corresponds with the pulse wave in the arteries. Therefore, he realized, the heart must be the cause of the arterial extension.

Next, Harvey carried out experiments which showed that the atria of the heart bear the same relationship to the ventricles as the ventricles do to the arteries. Using either the heart of a cold-blooded reptile or the dying heart of an exsanguinated mammal in order to see the rapid sequences, he satisfied himself that the whole cardiac apparatus is designed to move blood forward in one direction only. In addition, he saw that the valves at the root of the aorta and of the pulmonary artery are perfectly designed to prevent regurgitation. This had been already shown by several earlier observers, notably Leonardo, but, of course, his drawings were not available to Harvey.

Realdus Columbus had already established the pulmonary circulation and this was a key point in Harvey's hypothesis. He leaned heavily on Columbus and quoted him extensively when he finally published in 1628.

The most compelling single argument used by William Harvey was a quantitative one. If the

Michael Villanueva, known as Servetus, discovered the pulmonary circulation, but his work was lost to the world. Born in Navarre in about 1509, he succeeded Vesalius as prosector to Gunther and was an exemplary Galenist. In 1536, he met John Calvin, the religious reformer, and this turned him in a fatal direction. In 1553, he published the *Restitutio Christianismi* and sent a copy to Calvin. Mainly a critical theological treatise, it also contained his statement on the lesser circulation. It attracted the attention of the Inquisition who condemned Servetus and his work. He fled to Calvinist Geneva, but Calvin also found the book heretical and the tract and its author were burned at the stake on October 27th, 1553.

Andrea Cesalpino was born in the Tuscan town of Arezzo in 1519. He probably received his medical training at Pisa, where he later held the chair of anatomy. Before Harvey he used the word circulation and referred to *capillamento* as the termination of arteries and veins. Cesalpino was devoted to biology, botany, and mineralogy, and devised a system of classifying plants. He held the post of director of the botanical gardens before going to Rome as physician to Pope Clement VIII. Cesalpino was also an adherent of the school of philosophy that regarded all cosmic phenomena as the product of one single principle. Every being was considered to be a microcosm mirroring the macrocosm of the whole universe. This was in opposition to the prevailing Galenic theory of life, with its diverse "spirits."

William Harvey, who discovered the circulation, was the epitome of the new breed of scientists. A short, dark, and somewhat irritable man, he emerged from the Renaissance to free medicine from superstition and to revolutionize scientific research.

Born in Kent in 1578, Harvey was one of six sons of a prosperous farming family. At the age of sixteen he went to Caius College, Cambridge which had been recently founded by John Caius, the English anatomist who had attended the medical school at Padua. Harvey took an arts degree and then journeyed to Italy. During his four years at Padua, he was particularly influenced by his teachers Fabricius and Casserius. Harvey qualified at the age of twenty-four and returned home in 1602, at the beginning of the Stuart dynasty in England with which he was to be closely associated. He opened a practice in London, was admitted to the Royal College of Physicians in 1607, and two years later was appointed physician to St. Bartholomew's Hospital. There he was obliged to "attend the poor one day in the week throughout the year."

His appointment as Lumleian Lecturer at the Royal College in 1615 brought him great prestige. He lectured twice weekly, in six yearly cycles, for the next forty years, and had each year "to dissect all the body of a man for five days altogether, as well before as after dinner; if the bodies may last so long without annoy." The notes for these lectures are still in the British Museum. In 1625, Charles I came to the throne, and Harvey was appointed royal physician. Four years later he resigned his post as treasurer of the Royal College in order to travel with the young James Stuart on the continent. The portrait by Hannah below shows the anatomist explaining the circulation to the king and his son. Harvey was at his unfortunate sovereign's side through the Civil War, and he had charge of the young princes during the Battle of Edgehill. When he died, at the age of eighty, Harvey left an endowment to the Royal College for the Harveiian Oration, still delivered annually in his honour.

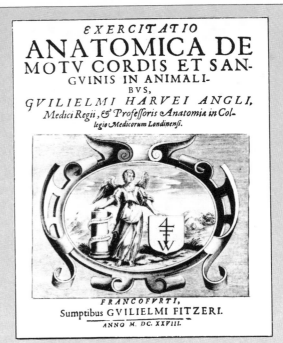

The frontispiece of Harvey's *De Motu Cordis*, first published as a slim monograph in Frankfurt in 1628.

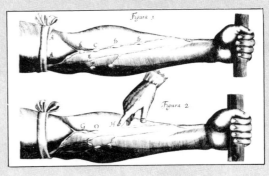

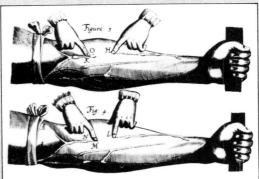

Harvey's experiments included this famous sequence featured in *De Motu Cordis*, on the tourniqueted arm of a living subject. By constricting the blood supply, he demonstrated that the direction of flow in the veins was always towards the heart, with the valves of the veins preventing back-flow. The plates were similar to a drawing published by Fabricius in 1603, showing valves in the veins.

heart can propel, say, two ounces of blood with each beat, then two ounces are ejected into the aorta, unable to return because of the valves. (The actual quantities were wrong, but that makes no difference to the validity of the argument.) Therefore, if there are seventy-two pulse beats per minute, the heart must be pumping no less than 8,640 ounces of blood every hour. This volume of fluid is equivalent to more than five hundred and thirty-two pounds—more than three times the weight of the average man. Harvey wondered, Where could such a mass of fluid come from? Where could it all go?

Columbus and Servetus had already answered the first question in their descriptions of the lesser, or pulmonary, circulation. To the second question, Harvey never had the answer, for that was to depend on the microscope. Forty years later, Marcello Malpighi—who was born in the same year that Harvey's book was first published—described the capillary system. But Harvey was content to leave it an open question, presuming that the blood either remained in vessels between the arteries and veins or that it somehow percolated through pores and cavities in the tissues.

The final step in Harvey's epic task of deduction appears to have been taken in 1615, the year before he delivered his historic lecture. He wrote in his notebook: "I began to think whether there might not be a movement, as it were, in a circle and afterwards I found this to be the case. I saw that the blood, forced by the action of the left ventricle into the arteries, was sent out to the body at large. In like manner it is sent to the lungs, impelled there by the right ventricle into the arterial vein."

Not content with mere hypotheses, William Harvey did many experiments on animals, especially snakes, where the heart and blood vessels were conveniently accessible. By blocking off vessels at various points, he proved that blood flow is in one direction only and that the pulsation and colour is dependent solely on the heartbeat.

Since he could hardly vivisect human beings, he performed his experiments on the surface veins of a living subject's arm. By pressing on filled veins, he showed that the blood flow was always in one direction—towards the heart. He saw that sections of the veins could be emptied at will by pressure, and that because of the valves, which had been so well demonstrated by his old teacher Fabricius, the empty segments of veins would always refill only from the peripheral end, when the pressure was relaxed.

It was not until 1628 that a newly published monograph, slender and rather badly printed, was seen at the Frankfurt Book Fair. The full title of this seventy-two page book, which was to preface a new era in medical and scientific methodology, was *Exercitatio Anatomica de Motu*

Cordis et Sanguinis in Animalibus (An Anatomical Treatise on the Movement of the Heart and Blood in Animals), but is almost universally known as *De Motu Cordis*. Dedicating it to King Charles I, Harvey compared the king to the heart, the centre of all strength and power.

Characteristically, Harvey admitted in the book that he had no idea why the blood circulates, only that it does. He suggested that the purpose of the circulation might be for the distribution of heat or for the sake of nourishment. He was right on both counts.

There is one extraordinary error in Harvey's book. He was so devoted to Galen, whose theory he was unknowingly in the process of discrediting forever, that he wrongly attributed to "that divine man, that Father of Physicians," knowledge of the lesser circulation. He misquotes a passage and indicates that Galen had said that the blood travels from the pulmonary artery to the pulmonary veins through the lungs. Galen, of course, had said no such thing, being the archetypal ebb-and-flow man.

Harvey was right, however, in prophesying that he should "tremble lest I have mankind at large for my enemies." The publication of *De Motu Cordis* brought a storm of criticism and invective upon him from ardent Galenists and his London practice declined considerably.

Of his opponents in the scientific world, probably the most virulent was Jean Riolan, a professor of anatomy and botany in Paris and one of those who stated that if anatomical findings at dissections did not agree with Galen, then the body in question must be abnormal. Riolan, who was a voluminous writer on

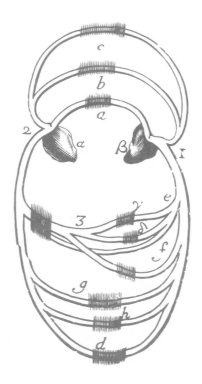

This diagram of the circulation, showing the capillary beds, is from a book published in 1671. It was drawn by Niels Stensen, the Danish anatomist who had been a staunch supporter of Harvey. The networks joining the arteries and veins were first observed ten years earlier by Morgagni.

Jean Riolan, who was born in 1577 and lived to the age of eighty, is one of the more ironic figures in the history of medicine. A cultured man who held the chairs of anatomy and of botany at the University of Paris, Riolan enjoyed a great reputation as a teacher and was a prolific writer, particularly on the circulatory system. He was nevertheless an ardent Galenist. He attributed to evolution—or to physical malformation—any contradictions of Galen's writings by scientific observers. When *De Motu Cordis* was published, Riolan leaped to the defence of Galen and launched a bitter and irrational attack on William Harvey.

anatomy with the reputation of being an outstanding teacher, was so fiercely critical of Harvey's discovery that the otherwise patient Englishman went to the trouble of addressing specific rebuttals, something he did not bother to do for his other adversaries. He wrote two disquisitions. In one, *De Circulatione Sanguinis*, published in 1649, he categorically refuted all Riolan's objections. Another harsh critic was James Primrose, a Scottish emigrant to France, who managed, within a fortnight of the publication of *De Motu Cordis*, to write and have printed a book condemning Harvey. Many of Harvey's detractors were from France, including Guy Patin, an important faculty member of the Paris school, who claimed the Englishman's work was "paradoxical, useless, false, impossible, absurd, and harmful."

Many more eminent men of the time came out in support of Harvey, however, and within a few years the tumult died down and he was permitted to enjoy the deserved credit of his great discovery.

After Harvey's brilliant exposition, discoveries and understanding of the more detailed anatomy of the heart seem something of an anticlimax. But a host of other men had much to contribute to the knowledge of cardio-vascular structure and function. Many of them left their names attached to various parts of the heart and circulatory system and this process continued into the twentieth century.

The actual fabric of the heart is rich in famous names. There is a small prominence on the wall of the right atrium, for example, which is known as Lower's tubercle. The name is more important than the structure, for it was Richard Lower, the English physician and a pioneer of blood transfusion, who showed in 1669 that the blood is oxygenated in the lungs. In a sense, his

work provided the final chapter in the Harvey story, since it confirmed the main purpose of the circulation of the blood.

The heart valves have some eponymous curiosities. The large semilunar pocket valves guarding the exits into the aorta and pulmonary artery have thickened nodules on the margins of their flaps, where the three cusps come together in the closed position. These are known as the corpora Aranzii, after an Italian anatomist Giulio Aranzi. Aranzius, as he is usually known, was a pupil of Vesalius at Padua. At the age of twenty-seven, he was appointed professor of medicine and surgery at Bologna and later became physician to Pope Gregory XIII.

Similar but smaller nodules are present on the flaps of the mitral valve (named after the bishop's mitre, the shape of which it resembles) and the tricuspid valve, which lie between the two atria and ventricles. These go by the name Albini's nodules. They were first described by Jean Cruveilhier, a pathologist who was active in the first part of the nineteenth century. Born at Limoges, France, in 1791, Cruveilhier lived to the age of eighty-two. In his time he saw many thousands of post-mortems, but he was so distressed at the first dissection he ever saw as a student that he ran away from medical school and entered a theological college. His father, who was a doctor, urged him to return to medicine, however, and Cruveilhier became one of the greatest of the early pathologists. He was professor of anatomy in Paris from 1825 to 1836 during which time he described the nodules on the heart valves. The Italian Giuseppe Albini, who wrote about the nodules in 1856, however, received the eponymous accolade.

At the base of the aorta, immediately behind the cusps of the aortic valve, are three bulges and from two of these the ostia of the coronary arteries emerge. These dilations are called the sinuses of Valsalva, named for yet another Italian anatomist, Antonio Maria Valsalva. Best known as one of the outstanding anatomists of the ear, Valsalva studied under Malpighi at Bologna, eventually succeeded him to the chair of anatomy there, and subsequently became the teacher of Giovanni Battista Morgagni. Much of Morgagni's fame as a pathological anatomist was founded on Valsalva's accurate anatomical descriptions, and Morgagni gave full credit to his mentor.

In the blood supply to the heart itself, a well-known name is that of Adam Christian Thebesius, a German who was an anatomist and pathologist at the University of Leyden in Holland. Thebesius made a special study of the coronary blood supply in 1708, and this led him to the discovery of the small veins, now bearing his name, which drain the blood from the muscular cavity of the right atrium. The actual mouths of these little veins are called Thebesian foramina, although some of the larger ones, described by the French pathologist Odillon Lannelongue in 1867, became known as Lannelongue's foramina. Thebesius's name is commemorated in the rather rudimentary valve lying at the mouth of the coronary sinus, the venous channel which drains most of the blood from the heart itself.

In the left atrium there is a small vein of no great importance except that it represents an embryological vestige (some animals other than man have a vena cava on the left side). This is sometimes called Marshall's vein after John Marshall, an English anatomist born in 1818. He is further commemorated in Marshall's fold, a ligament in the vena cava which is also of embryological interest.

Despite attempts to banish eponyms from anatomy, the circle of Willis is one of the hardiest survivors of all time. In 1664, Thomas Willis described the ring—or more accurately, the pentagon—of blood vessels that lies at the base of the brain and helps to ensure against any deprivation of blood supply to the brain.

Many arteries have names attached to them. A famous one in the brain is Charcot's artery, one of the anterior branches of the middle cerebral artery. It is named after Jean Martin Charcot, the nineteenth-century French neuropathologist and neurologist. Charcot's artery is also often known as the artery of cerebral haemorrhage because Charcot identified it as the vessel most frequently implicated in an apoplectic haemorrhage.

Guidi's artery, a small branch at the base of the skull, which passes through the pterygoid canal, is named for Guido Guidi. Better known by the Latinized name Vidius, he was born in Florence in 1500. He went to France to become professor of medicine in Paris and physician to King Francis I. He returned to Italy in 1548 and for twenty years was professor of philosophy and medicine at Pisa. Vidius was also bold enough to declare that Galen's pores in the septum of the heart did not exist and stated that "not a drop of blood could pass from right to left." His original and excellent writings were published by his nephew some years after his death in 1567.

Best known of all the eponymized veins in the skull are the veins of Galen. The great vein of Galen is the major vessel, which passes backwards from the centre of the brain; there are, too, smaller choroidal veins which are known as the veins of Galen. The middle cerebral vein is called Browning's vein, but it really should be attributed to the French anatomist Paulin Trollard who described it first in 1879. William Browning who described it five years later, was an American who studied in Leipzig, and returned to the United States to become professor of neurology at Long Island College Hospital.

Another vessel which drains the cerebral cortex is Labbé's vein, named after Charles Labbé, a nineteenth-century Parisian anatomist.

A small vein which passes out of the cranial cavity through the parietal foramen is called the vein of Santorini, and here again the man is much more important than the eponym. A number of structures bear his name, for he was one of the leading Italian anatomists of his day.

Associated with the cerebral veins are various sinuses, venous spaces built into the layers of the dura mater, the outermost membrane of the brain. At the point where several sinuses meet within the tough membranes on the inside of the back point of the skull, there is a slightly twisted dilation, called the confluence of the sinuses. This has long been known as the torculus Herophili, the wine-press of Herophilus, recalling the famous Alexandrian anatomist. A much more modest blood sinus is the petro-occipital,

Thomas Willis, an outstanding physician of the seventeenth century, an iatrochemist, and a professor of natural philosophy at Oxford, was born in Wiltshire in 1621. A member of the circle of extraordinary men who founded the Royal Society in 1660, Willis made a study of many of the diseases which swept Europe, especially influenza and typhoid fever. He identified childbed fever, which he named puerperal fever, recognized that asthma is caused by constriction of the bronchioles, and pointed out the presence of sugar in the urine of diabetics—which can be ascertained by tasting the sample. In 1664, he published the most complete and accurate account to date of the nervous system, and described the condition of myasthenia gravis, a progressive paralysis. Willis died in London at the age of fifty-four.

Jean Martin Charcot, the famous neurologist and neuropathologist, was born in Paris in 1825. At the hospital at Salpetrière he established the greatest clinic of the time for neurological diseases. Charcot, who was said to be an incomparable writer, teacher, and leader, was the first to describe many of the most important nervous diseases, including hysteria, multiple sclerosis, and muscular atrophy. As a clinician, he influenced Freud's thinking. Charcot established neurology as a separate discipline of medicine and, in 1882, eleven years before his death, a chair in nervous diseases was created for him.

sometimes called Englisch's sinus after the Viennese physician Josef Englisch, who described it in 1863.

Throughout the body there are blood vessels which bear men's names. In the liver, the central veins of the lobules are known as Krukenberg's or Kiernan's veins. Adolf Krukenberg, professor of anatomy first at Brunswick and later at Halle, seems to have a weaker claim since he published in 1843. The other contender was an Irishman, Francis Kiernan. A lecturer in anatomy at London's Royal College of Surgeons, he was a pioneer in discovering the fine structure of the liver and wrote his account of the central veins in 1833.

There are also rival claims for the pre-pyloric vein at the exit of the stomach. This was originally described by André Latarget, professor of anatomy at the University of Lyon, who was born in Dijon in 1877. The vein which accurately indicates the position of the pylorus (the muscular ring controlling the outflow from the stomach) is, however, frequently called Mayo's vein, commemorating the famous Mayo brothers of Rochester, Minnesota—the founders, together with their father, of the Mayo Clinic—who mentioned the vein between 1906 and 1913.

One of the longest names in anatomical eponymy must surely be that of Vikentiz Bonifatiyevevich Konstantinowich, a surgeon and anatomist of St. Petersburg, where, in 1872, he published an account of the various blood vessels around the anus. The context was mainly surgical and his surname has since been attached to the marginal vein of the anus and to a branch of the superior rectal artery.

Probably the most eponymized doctor in the United States was Dr. Frederick Byron Robinson—to such an extent that in 1905 the American Medical Compendium published a list of Robinsonian terms. As far as the circulatory system is concerned, he is remembered for Robinson's circle, a ring of arteries in the uterine and ovarian region, and for Robinson's cervical loop, part of the uterine artery adjacent to the neck of the womb.

Frederick Byron Robinson was born in Hallondale, Wisconsin, in April 1855 and graduated from Rush Medical College in 1882. He practised in Grand Rapids for some time before going to work and study in Europe. He returned in 1889 to become professor of anatomy at Toledo Medical College and soon moved to a similar post at the University of Chicago. He died in River Forest in 1910, having published extensively. His best known work was *The Peritoneum.*

While cardiac physiologists had known for a long time that heart muscle has the power to contract independently of outside stimulation, it is only fairly recently that medical men began to see that, in order to function efficiently in different circumstances, the heart requires control and co-ordination from outside itself. The various parts need sequential contraction, so that the atria contract first and then the ventricles contract in a way which will provide the most effective ejection of blood through the semi-lunar valves. This co-ordinated rippling movement is mediated, or controlled, by a communications system in the heart and also by nerve control outside the heart which modifies the heart-rate according to varying demands of exertion, rest, and emotion.

The internal system is mediated by specially modified muscle strands. The whole contractile mass of the heart is formed of an interlacing web of muscle fibres, which, unlike the skeletal muscle, all communicate with each other. Some of these fibres take on the special role of conducting the electrical impulses which fire off the contraction. Cardiac-conducting fibres are larger than normal muscle fibres, are more primitive in structure, and have clearer cell contents when seen under the microscope. Most run immediately under the internal lining of the chambers where they were discovered by the outstanding Bohemian physiologist, Johannes Evangelista Purkinje in 1839.

The fibres which Purkinje saw, and which are now universally known as Purkinje fibres, run in two main branches passing down each side of the interventricular septum, the wall between the ventricles. The one on the left penetrates the septum near its upper end, but the branch serving the right ventricle passes down the wall, some of it crossing within the moderator band, which had been drawn by Leonardo in the fifteenth century.

The main trunk, or bundle of fibres, from which these two branches arise was named simultaneously—but independently—by Wilhelm His the Younger and the English physiologist Albert Kent.

Giovanni Domenico Santorini was born in Venice in 1681 and had the distinction of studying in all three of Italy's hallowed centres of medical learning—Bologna, Padua, and Pisa. In 1703, he became professor of anatomy and medicine in Venice. He was the author of a number of books, one of which, *Observationes Anatomicae,* published in 1724, was dedicated to Peter the Great of Russia. His anatomical illustrations, published posthumously in 1765—he died in Venice in 1737—are said to be among the medical masterpieces of the eighteenth century.

The famous son of a famous father, Wilhelm His was professor of anatomy successively at Leipzig, Basel, Göttingen, and Berlin, and is well known for his important studies on the rhythmicity of the heart. Albert Kent, who was also an early pioneer in X-rays, was born in the same year, 1863, as Wilhelm junior and published his account of the atrio-ventricular bundle in 1893, the same year as the German anatomist. So, although the usual name for this vital structure in the cardiac conducting system is the bundle of His, the term Kent's bundle is equally justified.

The atrio-ventricular bundle, by whichever name it is called, arises from a nodule of conducting tissue situated between the atria and the ventricles, embedded in the wall of the right atrium. This nodule, anatomically known as the atrio-ventricular node, is also called the Tawara and Aschoff node. These names are applied in tandem, rather than as alternatives, since the two men worked together. K. Sunao Tawara was a Japanese anatomist and pathologist, born in 1873, who came to Europe and worked with Professor Ludwig Aschoff, the renowned German pathologist, at the University of Marburg. It was Tawara who, in 1906, published an account of their work on the conducting tissues of the heart and described the atrio-ventricular node—the node of Tawara and Aschoff—before returning to Japan.

Almost immediately after this, another important node was discovered, this time situated higher up in the heart. This was the sino-atrial node, the heart's natural pacemaker. It sets the rhythm of the heart by discharging electrical impulses down the connecting fibres to the node of Tawara and Aschoff, which passes then, by way of the bundle of His, to the ventricles. This structure was described by an eminent Scot, Sir Arthur Keith, and an Englishman, Martin Flack. Their findings on the sino-atrial node were published in 1907. Once again a double eponym was invoked and the cardiac pacemaker became known as the node of Keith and Flack.

A little-known competitor to Keith and Flack is the German Walter Koch, who was responsible for describing in 1912 the area of the wall of the right atrium containing the atrio-ventricular node which is now known as Koch's triangle. German anatomists also associated his name with the *sinusknoten*, or sino-atrial node, described by him in 1909. This was, however, two years after its description by Keith and Flack, so their claim to priority seems secure.

It is extraordinary how medical research often proceeds on parallel lines. Both His and Kent made their discovery independently in the same year, and Keith and Flack together and Koch, working alone, found the same tiny, but vital, nodules within two years of each other. These

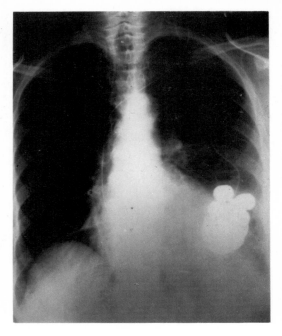

A life-saving invention for some heart patients, the artificial pacemaker shows up clearly, beneath the shadow of the left lung, on an X-ray film. When the heart's natural pacemaker, the small mass of nervous tissue called the sino-atrial node, is damaged by disease, its impulses cannot reach all parts of the heart, causing an irregular beat, or heart failure. The cardiac pacemaker implant is a battery-like unit which provides the impulse to stimulate the diseased heart, allowing it to beat at normal rate.

coincidences stem partly from related fields of interest, where the work is stimulated by other discoveries. But sometimes, too, coincidental discoveries can lead to bitter disputes about priority of claims and even to plagiarism.

Various other structures in the heart, separate from the conducting system, are related to rhythm and nervous control. A ganglion, for example, is a collection of nerve cells, similar to a junction box on an electrical circuit, which are continuous with the autonomic nervous system. The two types of tissue—nervous and muscular—interact to allow some nervous control over the heartbeat. Two German anatomists associated with heart ganglia are Bidder and von Bezold. In 1836, Friedrich Bidder found his ganglion situated in the septum, between the two atria. Also between the two atria, but higher up in the septum is Bezold's ganglion, described by Albert von Bezold in 1863.

In the previous year, von Bezold had shown that there are accelerator nerve fibres in the heart, which he traced to their source in the spinal cord. Bezold's accelerator nerves are also known as Gaskell's nerves, after Walter Holbrook Gaskell, a lecturer in physiology at Cambridge, who described these same nerves in 1881.

It is when any of these vital cardiac structures, especially the bundles, are damaged by disease, that an artificial pacemaker is now implanted and provides a replacement of the lost electrical impulses which serve to trigger the heart at a faster rate than the slow beat which is all that the intrinsic rhythm will allow. The work of these later men, therefore, in elucidating the conducting mechanism of the heart has been of fundamental importance in the treatment of many common types of heart disease.

The Respiratory System

With their long tradition of embalming, the early Egyptians were familiar with the organs of the chest cavity. Among the stylized anatomical drawings which passed into their elaborate system of hieroglyphs were symbols, shown above, representing the trachea and lungs.

In the early history of anatomy, there is almost total silence about the air passages and the lungs, even though it must have been obvious to ancient observers that the rhythmical movements of the chest and the passage of air through the nose and the mouth were absolutely essential to life. It is true that the wind that could be felt and heard in breathing was less tangible than the solids and liquids of nourishment, but an appreciation of gales and storms and draughts should have created analogies in their minds.

Despite the obvious, respiratory functions figured only slightly in the fantastic schemes conjured up by the early medical theorists of Greece and Rome. Although that wondrous substance pneuma was closely related to the air and vapours of the universe, the function of the lungs and air passages was treated in a perfunctory manner when it came to explaining the formation of the various natural spirits and other intangible substances that formed the basis of Galen's scheme of physiology.

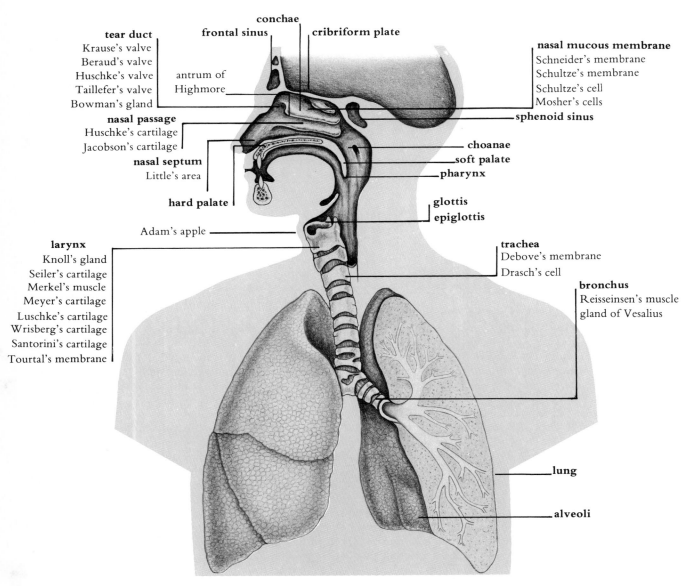

tear duct
Krause's valve
Beraud's valve
Huschke's valve
Taillefer's valve
Bowman's gland

antrum of Highmore

conchae
frontal sinus

cribriform plate

nasal mucous membrane
Schneider's membrane
Schultze's membrane
Schultze's cell
Mosher's cells

sphenoid sinus

nasal passage
Huschke's cartilage
Jacobson's cartilage

nasal septum
Little's area

hard palate

choanae
soft palate
pharynx

glottis
epiglottis

Adam's apple

larynx
Knoll's gland
Seiler's cartilage
Merkel's muscle
Meyer's cartilage
Luschke's cartilage
Wrisberg's cartilage
Santorini's cartilage
Tourtal's membrane

trachea
Debove's membrane
Drasch's cell

bronchus
Reisseinsen's muscle
gland of Vesalius

lung

alveoli

Broadly, his concept was that pneuma from the atmosphere passed down the air passages and made direct contact, through the blood vessels, with the ventricles of the heart, vitalizing the blood and removing waste matter. If oxygen is substituted for the vitalizing process and carbon dioxide for the waste products, then the scheme is almost acceptable, but the ancient hypotheses in their convoluted entirety were really too outrageous to make it possible to use such an analogy.

According to Plato and Aristotle, the very core of the life process was the innate heat produced by a fire burning in the heart. This resulted in the warmth upon which depended all vital processes. This *flamma vitalis* was said to be fuelled by the nourishment of the body. Breathing moderated it by keeping it cool and preventing it from consuming itself. Just as the brain was once considered merely a radiator for the heart, so the bellows of the chest were assumed to be a sophisticated type of fan to regulate the vital flame.

As for the anatomy of the respiratory system, the Egyptians accumulated some crude knowledge of the organs in the chest. Stylized representations of the heart, trachea, and lungs were used as amulets and some even passed into the Egyptian hieroglyphic system of writing. And votive models used for divination in Roman times show fairly recognizable details of the rib cage and chest organs. But until the Middle Ages the respiratory machinery was largely ignored.

The respiratory system consists basically of the lungs and the passages which connect the lungs to the outside of the body. These passages have the accessory function of producing the voice in the larynx and the sense of smell high up in the nasal passages. The top end of this airway is shared by the alimentary canal. The parting of the way is at the level of the larynx and pharynx, where the glottis forms the entrance to the air tubes, protected by the flap of the epiglottis, which diverts food down into the upper end of the oesophagus.

No real progress could be made in understanding the purpose and function of the respiratory apparatus until the circulation of the blood was worked out, since the two systems are structurally and functionally related.

Michael Servetus, the martyred Spaniard, was the first to declare, in 1553, that blood passed through the lungs from the right to the left side of the heart. Then, in 1558, the Italian Realdus Columbus, stated: "The blood goes through the *vena arteriosa* (pulmonary artery) and there is attenuated. Then, mixed with air, it goes through the *arteria venosa* (pulmonary vein) to the left heart." Towards the end of the sixteenth century, Andrea Cesalpino, another Italian, stated that the blood that reaches the lungs from

the heart must be distributed in fine branches and come into contact with the air that penetrates from the finest air passages. He also corrected the Galenic view that blood and air come into contact by mixing directly together. It was obvious to Cesalpino that a fine membrane must separate them, as indeed it does in the diaphanous wall of the lung capillaries.

The main organs of respiration, the lower air passages and the lungs, are unusual in that they are almost totally devoid of eponyms. The discoverers and describers of their anatomical details are virtually anonymous, in sharp contrast to other systems. And even the few structures that are named for their describers are not great in magnitude. Debove's membrane, for example, is the thin layer of cells under the lining of the main air passages. The actual lining of the trachea and bronchi consists of cells with an inner border of cilia, microscopic whip-like hairs which beat incessantly in the direction of the throat, wafting out mucus containing entrapped dust, debris, and harmful bacteria. The foundation for this delicate layer of cells is named after Georges Maurice Debove, a Parisian pathologist who lived from 1845 to 1920.

In the same area are Drasch's cells, the cuneiform, or wedge-shaped, cells of the mucous membrane of the trachea. They were first described in 1886 by Otto Drasch, an Austrian who was professor of histology at Graz.

Outside the mucous membrane are layers of smooth muscle fibres which narrow the air passages in certain circumstances, including the broncho-spasm of asthma. In the smallest bronchi, the layer is called Reisseinsen's muscle, after François Daniel Reisseinsen, an anatomist and physician of Strasburg, who also practised in Berlin, where he died in 1828.

The only other eponymous structure in the lower air passages bears the name of Vesalius. The small lymphatic glands associated with the bronchial tubes are called the glands of Vesalius and are among the few rather insignificant, and even inconstant, structures that remain to perpetuate the name of the great Reformer of Anatomy.

The larynx, which is basically a box of cartilage, or gristle, containing muscles and ligaments, abounds with eponyms. Karl Ludwig Merkel described in 1837 two structures in the larynx—the cerato-cricoid muscle and the fossa between the two ventricles of the larynx—and they were named after him. A professional larynx man, he was professor of laryngology at the University of Leipzig in Germany.

Cartilages provide niches for other names in laryngeal anatomy. Giovanni Santorini, the distinguished eighteenth-century Italian anatomist, who is eponymized in the pancreatic duct and the veins, also has his name attached to the

Prior to the Middle Ages the mechanisms of respiration were ignored, nevertheless the medieval Church created a patron saint of the lungs—Saint Bernardine of Siena. A Franciscan who lived from 1380 to 1444, Bernadine devoted his life to the care of the sick.

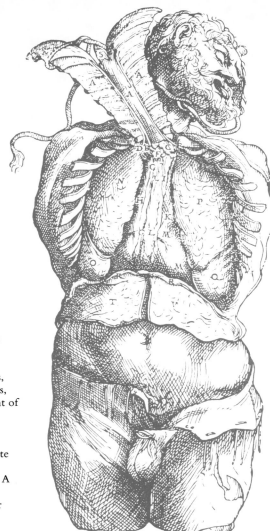

Vesalius dealt with the lungs only briefly—in the sixth book of his *De Fabrica*. This illustration shows the body with the ribs cut through and the sternum raised to reveal the underlying structures, including the heart, lungs, and thymus. Reminiscent of the use of the bodies of criminals for dissection, the presence of the rope may, however, only relate to its use as an aid to post-mortem technique. A curious affectation is the clothing of the lower limbs in britches and codpiece.

a professor at Göttingen from 1764 until his death in 1808.

Another German professor of anatomy also has an impressive list of things associated with his name, including yet another laryngeal cartilage. This was Hubert Luschke, who described a number of small accessory cartilages, including the inter-arytenoid. He was professor at Tübingen between 1849 and 1875.

There are small glands situated in the false vocal cords, ridges of membrane above the true cords which produce the voice. These glands have been the subject of some controversy, since many anatomists denied their very existence. It now seems firmly accepted that they are not the figment of the imagination of Philip Knoll, professor of experimental physiology in Prague and then Vienna, who described Knoll's glands in 1873.

The ligaments which join the arytenoid cartilages with the epiglottis are known as Tourtal's membrane. A lecturer in anatomy, Kaspar Tourtal was born, worked, and died, in 1866, in Münster, Germany.

Undoubtedly the best known eponym in the human body is the Adam's apple, the prominence visible in the front of the neck. More evident in men than in women, and caused by the thyroid cartilage, it is so called because of the legend that a piece of the forbidden fruit stuck in Adam's throat.

The nose is part of the respiratory system. One of its functions is to accommodate the organ of smell. This is necessarily combined with the function of inspiring air, since it is the atmosphere that carries those few molecules that so exquisitely stimulate the olfactory nerve endings.

In humans the ability to detect odorous substances in low concentration is rudimentary, compared with many animals. In some species, including insects, this sense is fantastically developed, so that some varieties of moth, for example, can detect the few molecules emitted by a female at several miles distance. In many of the other animal species, the part of the brain which deals with the input from the smell sense organs is relatively enormous, sometimes dominating the rest of the cerebral hemisphere. In humans, this rhinencephalon, or smell-brain, is relatively tiny.

The nerve endings which pick up the molecules in the air are situated in the upper part of the complex nasal passages, where nerve fibres penetrate downwards from the brain, through the bony floor of the front part of the base of the skull. This perforated bone is called the cribriform plate from the Latin word for sieve.

The air enters through the two nostrils and passes backwards to reach the back of the throat through the rear apertures, the choanae, meaning funnels. Plates of thin bone project from the

corniculate cartilages, the arytenoids, which hang over the entrance to the larynx, and is remembered in some of the internal bones in the nasal passages.

Another tiny cartilaginous structure attached to the arytenoids, to which the vocal cords are also attached, is known as Seiler's cartilage. Carl Seiler was a Swiss laryngologist who studied in Germany and Austria before going to America where he qualified in Philadelphia in 1872.

A sesamoid cartilage or bone is one which appears free-standing in a tendon or ligament, such as the kneecap. Meyer's cartilage is a loose sesamoid nodule suspended in the front edge of the lower thyro-arytenoid ligament of the larynx. It is named after Edmund Victor Meyer, who became professor of disorders of the ears, nose, and throat in Berlin in 1902.

In the larynx, Wrisberg's cartilages are the cuneiform structures often present in the front of the voice box. They are named for Heinrich August Wrisberg, a German anatomist who was walls of the nasal passages, causing turbulence and offering a larger surface area. This warms the inspired air before it enters the tubes and lungs and also gives the moist membranes of the nasal passages a better opportunity to filter out much of the dust and bacteria, which the ciliated membrane then wafts back towards the nostrils. These plates are called the conchae, from the Latin for shells, because the convolutions resemble a mollusc's spiral.

The design of the nasal air tunnels is such that when rapid, short breaths are taken in—in other words, sniffed—the turbulence drives air up to the top of the domed passages where the smell nerves are placed, so facilitating the detection of any odours.

The nasal mucous membrane has several eponyms attached to it. It has been known as Schultze's membrane and its constituent cells as Schultze's cells after the nineteenth-century German anatomist Maximilian Johann Sigismund Schultze.

Much earlier, the part of the nasal membrane which lines the roof of the cavities where the smell sense lies, was given the name Schneider's membrane. Conrad Viktor Schneider, professor of medicine in Wittenberg, was one of the first men to discover the exact location of the olfactory nerve endings and wrote an adequate description of them in 1655.

The same nasal lining, but lower down in the septum that divides one nostril from the other, may be the site of profuse nosebleeding, especially in persons with high blood pressure. This zone of nosebleeding is called Kiesselbach's area after Wilhelm Kiesselbach, a professor of otology at Erlangen, Germany. He drew attention to this danger area in 1884, in an article on "Spontaneous Nosebleeding." Another surgeon described the same general area on the nasal septum, but this time in relation to a cancerous tumour. Little's area was named after James Laurence Little, who was born in Brooklyn, New York and became a professor of surgery in Vermont.

A cartilage, known as the vomero-nasal cartilage, forms part of the wall of the nasal passage and is called either Jacobson's cartilage or Huschke's cartilage. Ludwig Jacobson, a physician, was born and died in Copenhagen between 1783 and 1843. His name is attached to a number of structures in the nose and ear. Emil Huschke was an almost exact contemporary of Jacobson, but was born in Weimar and was a professor of anatomy at Jena, Germany.

The large air spaces in the bones of the face and skull are called the para-nasal sinuses and occupy much of the skull between the upper jaw and the forehead. They add to the resonating properties of the voice, but their main function seems to be the lightening of the mass of the skull without

sacrificing any strength. The largest ones, the maxillary antrum in the upper jaw and the frontal sinuses in the forehead, were drawn by Leonardo da Vinci early in the sixteenth century.

The maxillary sinus is known as the antrum of Highmore, although Nathaniel Highmore could lay no claim to being the first to notice this antrum, or cave.

The frontal sinuses are interesting because they can be used in X-ray identification of the living or dead; no two persons in the world have exactly similar outlines to these air spaces. They are as individual as fingerprints and usually survive the most severe injuries.

Deeper in the skull are the smaller ethmoid and sphenoid sinuses, ethmoid meaning sieve because of numerous associated holes in their walls, and sphenoid being taken from the Greek for wedge-shaped. An extension of the ethmoid sinus is sometimes known as Mosher's

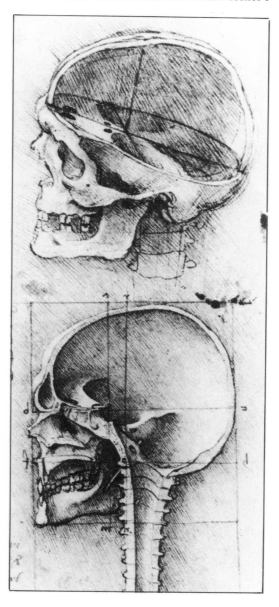

The para-nasal sinuses—cavities between the top jaw and the forehead—enhance vocal resonance and lighten the bony mass of the skull. Leonardo da Vinci drew the largest of these sinuses, the maxillary antrum in the upper jaw, and the frontal sinus in the forehead, in perfect detail. Also ahead of his time was his use of the cutaway wedge device, seen in the top drawing, to show the inside of the skull, including some of the sinus cavities.

Nathaniel Highmore, the English physician and naturalist, was born in Hampshire in 1613, qualified at Oxford in 1641, and went into practice in Dorset. He discovered the antrum which bears his name when pulling a tooth, presumably a marginal task of a country doctor. He could not have been an ordinary general practitioner, however, since he wrote a Latin text of anatomy, published in 1651, in which he described the maxillary sinus, a structure known to early anatomists and clearly illustrated by Leonardo da Vinci. Highmore, who died in 1685, described the mediastinum testis, formerly called Highmore's body.

Richard Lower was a Cornishman, born at Tremere in 1631. He read medicine at Oxford where he was a pupil of Thomas Willis, from whom he derived his own interest in the circulatory system and was influenced by the "new philosophy" that was questioning the reverence for Aristotle and Galen and challenging the hold of scholastic philosophy on science. When Lower went to London, he met other members of the Royal Society, including Boyle and Hooke, and in 1667 he was elected a fellow of the society. A pioneer of blood transfusion, Lower died in London at the age of eighty.

Robert Boyle, an aristocrat of Irish family, was not a doctor but an affluent gentleman with great scientific curiosity. Born in 1627, he attended Eton, then travelled with a tutor through Europe where he met Galileo. Independently wealthy, he was able to devote his life to physics and chemistry. Boyle was a prominent member of the "Invisible College," a group of young men who later became the Royal Society for Improving Natural History. Boyle's health was always poor, but when he died in 1691 he left behind numerous publications on all aspects of science, particularly chemistry and the physical properties of gases.

cells, after Harris Peyton Mosher, professor of laryngology at Harvard and chief surgeon at the Massachusetts General Hospital in the early part of the twentieth century.

There is sometimes a space within the ethmoid bone, communicating with the other sinuses. This is called Palfryn's sinus. It is named for Jean Palfryn who was born in Courtrai in Flanders in 1650 and was professor of anatomy and surgery in Ghent and later in Paris.

The tears which come from the lachrymal glands in the eyelids drain away through two tiny apertures in the inner corner of the eye. They pass into the tear ducts, the naso-lachrimal ducts, and end inside the nose — which is why snivelling accompanies weeping. These ducts have small valves which have attracted a number of different eponyms. They are either called Beraud's valves, after Bruno Beraud, a Parisian anatomist who described them in 1854, or Krause's valves, after Wilhelm Krause, another anatomist from the University of Göttingen.

The tear duct also has Taillefer's valves, named after a doctor who rejoiced in the name of Louis Auguste Horace Sydney Taillefer and described it in 1826. Last, there is Huschke's valve, the edge of the opening of the tear duct into a blind sac at the top of the duct. This Huschke is the same as the one whose name, along with Jacobson, is attached to the vomeronasal cartilage.

Finally, an English knight is commemorated in the anatomy of the nose, as he is in the kidney and the eye. This is Sir William Bowman, who described glands in the mucous membrane of the nose, later known as Bowman's glands.

When in the seventeenth century William Harvey's full explanation of the entire greater and lesser circulation was accepted, the way lay open for investigation into the process of respiration. Some of the work was done by anatomists, but much of it was carried out by chemists who began to delve into the nature and properties of air and gases.

Harvey himself had questions about the purpose of the passage of blood through the lungs. He had difficulty shaking off the conditioned reflexes of thought which his Galenist education had impressed upon him. At first he believed that the lungs were indeed a cooling system and that the air tempered the hot blood and prevented excessive bubbling. But to his astute mind this was a facile and unsatisfactory explanation. The true function of pulmonary circulation bothered him for the rest of his life. In his lecture notes made after the publication of *De Motu Cordis*, he remarked that "there is no life without breathing and no breathing without life." The question — why both a flame and an animal need air to survive — that exercised the contemporary Oxford scientists, also gave him much thought.

He intended to write on the subject, but never did so. He did eventually retract his Galenist statements, however, but only to deny that air is necessary either as a cooling medium or a nutrient. Harvey did not have the necessary later technology and chemical knowledge to do justice to his extraordinary perceptiveness.

The fine structure of the lungs was first observed by Marcello Malpighi in 1660, three years after Harvey's death. He described it in two letters to his friend and teacher, the mathematician Gian Borelli. In his first letter Malpighi described the microscopic appearance of the terminal air sacs, or alveoli, in the dog. In his second letter he told Borelli how he had seen the flow of blood through the capillary vessels adjacent to the alveoli in a living frog's lung. A year later Malpighi's findings were published in his book *De Pulmonibus.*

Malpighi was the first man to see lung circulation, but he did not understand the phenomenon. While he did not believe that its purpose was to cool the blood, he thought that the blood might be mixed and stored in the lungs, which acted as a kind of reservoir. Malpighi's observations of the capillaries was the final vindication of Harvey's theory of circulation, for the Englishman had had no idea how the circuit was completed from arteries to veins.

Gian Alfonso Borelli was a member of the iatrophysical, or iatromechanical, school, one of the medical cults that sprang up in the seventeenth century. Borelli had a laboratory in his own home in Pisa and there he worked, applying the laws of mechanics and statics to all physical phenomena.

A professor of mathematics, Borelli was equally interested in anatomy and all things biological. As the result of his talents and great energy, he made the University of Pisa famous as a school of mathematical and medical sciences. Although Borelli had vast intellectual gifts, he had an unpleasant personality. Quarrelsome and morose, he was jealous of his own position and envious of the success of others. Many of his colleagues thought him unbearable. Malpighi, his student and later his friend, looked upon him as a trusted confidant and almost like a father. But they became estranged, too, when Borelli opposed and criticized Malpighi's views—although Borelli's own work was built on the foundation of Malpighi's findings.

Borelli did much work on the movements of breathing. He denied the Galenic theory that breathing ventilated the vital flame in the heart and realized that "so great a machinery must have been instituted for some grand purpose." Borelli regarded the chest as a complex mechanism and attained a fairly good idea of how the chest cage and diaphragm caused air to enter the lungs. The diaphragm, Borelli pointed out, acts like a piston in a pump. When it moves down,

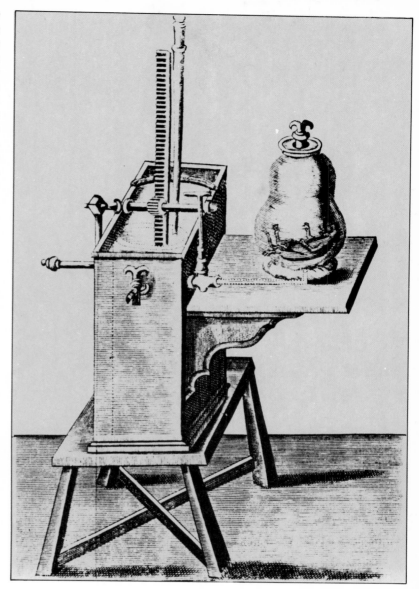

due to muscular contraction, it increases the volume of the chest cage and any distensible bag, such as the lung, which is connected to the outer air, has to increase in volume as the result of the decrease in pressure inside the chest. A similar increase in volume is caused by the ribs moving upwards and forwards, under the action of all the intercostal, or between-the-ribs, muscles. Borelli's work on the movements of respiration was a useful advance, although some of his hypotheses were fallacious.

The next major step in the understanding of respiration was taken at Oxford University in England, where a group of young men contributed greatly to the physiology of breathing. The first was Robert Boyle, who is celebrated in physics in Boyle's law of gases, familiar today to every English-speaking schoolchild. (In France, the same facts are known as Mariotte's law because it was independently discovered in

Robert Boyle, a pioneer of physics in the seventeenth century, used an air pump to demonstrate the effect of lack of air on an animal imprisoned in a bell jar.

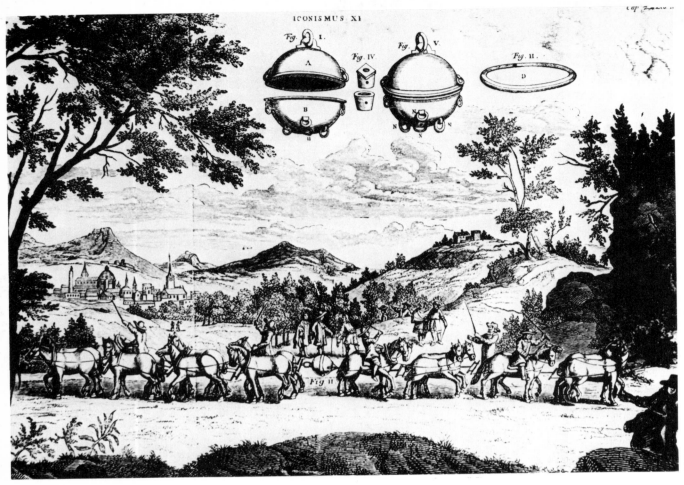

This seventeenth-century print shows Guericke's experiment which proved that once a vacuum is created inside a closed vessel its halves cannot be prised apart.

1676, fourteen years after Boyle, by a French physicist named Mariotte.)

Although Boyle's interest was in the behaviour of gases, he extended this to their effect on animals. His pupil Robert Hooke became his assistant and colleague and together they improved the newfangled air pump invented in

Germany by Otto Guericke, the Burgomeister of Magdeburg. It was Guericke who, in 1654, performed the famous experiment showing that teams of horses could not separate the two halves of his Magdeburg hemispheres once he had used his pump to create a vacuum inside them.

Using the air pump, Boyle proved that a candle could not burn, nor could a small bird or mouse live, inside a jar from which he had exhausted most of the air. If air was readmitted soon after the animal collapsed, it would often recover. It was therefore obvious that air was absolutely necessary for combustion and life.

On October 24, 1667, Robert Hooke performed the well-known, if barbaric, experiment, in which he opened the chest of a dog and found that it could be kept alive if air was forced into its trachea with a bellows. Vesalius had done much the same experiment well over a century before, showing that an animal with collapsed lungs could be revived by blowing into a reed tied into the windpipe.

Hooke's experiment proved that the essential function of the lungs is respiration, because when the artificial respiration was stopped the dog died. This was the first recorded use of such artificial respiration, the direct forerunner of the iron lung and of more sophisticated modern

Joseph Black, born in Bordeaux in 1728, went to Scotland for his studies. He read medicine and natural science at Glasgow and then completed his education at the famous medical school in Edinburgh. Black, who became a professor of chemistry at Glasgow in 1756, anticipated Lavoisier's discoveries in chemistry when he identified carbon dioxide as a separate gas in the atmosphere and noted that it is produced by fermentation and in respiration. He also discovered the bicarbonates, developed the concept of latent heat, and made many observations on heat, fusion, and vaporization. Black died in 1799.

mechanical life-support equipment. In the *Old Testament*, however, there is the description of the prophet Elisha bringing back to life the boy of the Shunamite, by blowing into his mouth. This, obviously, is the first example of mouth to mouth resuscitation.

Richard Lower, the London anatomist and physician, was interested in Boyle's findings for he was trying to discover why the blood in veins and the blood in arteries are of different colours. He observed that a blood clot in a long glass tube turned bright red at the end exposed to the air. When he turned the tube around, the clot at the other open end lost its bluish tinge and became bright red. He repeated Hooke's experiment with a dog and found that the blood issuing from the lungs became red when the bellows were worked, but subsided to a blue colour when artificial respiration was stopped. He also found that tying off the windpipe caused the blood to darken and that even when breathing is taking place, the blood from the right side of the heart is much darker than from the left. All this he described in his book *Tractatus de Corde*, published in 1669.

Lower concluded that it is the admixture with air in the lungs that alters the blood's colour, concluding that "it is easy to imagine the great advantage accruing to the blood from the admixture of air and the great importance attaching to the air always being healthy and pure . . . wherever, in a word, a fire can burn sufficiently well, there we can equally well breathe."

John Mayow worked at Oxford in the second half of the seventeenth century. Although he lived only to the age of thirty-four, his research was as brilliant as it was important. He was a lawyer before he became a physician, which might explain his clarity of thought and meticulous records.

Mayow collected all that was known about respiration and added his own original findings. He fitted a bladder inside a pair of bellows and confirmed Borelli's concept that the chest movements operated by reducing the pressure outside an elastic organ. When the bellows were opened, the bladder inflated through the nozzle and when the handles were squeezed, the bladder subsided. He again pointed out the part played by the diaphragm and by the muscles between the ribs and showed that although inspiration was an active movement of the muscles, expiration was a passive process due to the elasticity of the lungs and the relaxation of the muscles.

Mayow then turned his attention to the changes that went on in the lungs during breathing. He realized that something which was vital to life, something from the air, entered the blood. He then discovered that when either a candle or a small animal was placed on a platform in a bowl of water and covered by a bell

Concerned both with the mechanics and chemistry of respiration, John Mayow devised an experiment to prove that the volume of air in a confined space is reduced by breathing. He put a mouse in a glass jar, with a moist bladder stretched across to cover the entrance. As the mouse breathed, gradually the fabric of the bladder was drawn up into the glass, in place of the spent air.

Robert Hooke was a remarkable polymath who suffered from an irritable temperament, was physically misshapen and unkempt. No portrait of him exists, perhaps because of his unfortunate appearance. He was born on the Isle of Wight in 1635. A clergyman's son, he intended to follow his father's profession and began to study theology at Oxford. Ill-health forced him to abandon the ministry and he took up "natural philosophy," as science was called. Hooke became an employee of Robert Boyle the chemist, and he probably discovered the principle behind Boyle's law. In 1662, Hooke joined the group of philosophers and scientists in London who were at the centre of the scientific Renaissance. He was appointed Curator of Experiments to the Royal Society and was elected a fellow the following year. At the age of twenty-nine, he became professor of geometry at Gresham College where he was a colleague of Christopher Wren. Hooke's list of scientific achievements was prodigious, although it is said that they would have been more striking if they had been less diverse, since they ranged from physics to microscopy, geology, chemistry, meteorology, and metallurgy.

The first person to use the compound microscope for scientific investigation, he proposed a theory of evolution based on the fossils he observed through his lenses. Hooke's law of stress was the basis for the understanding of elastic materials, and with it he invented a spiral watchspring. His diagrams of Mars were used in the nineteenth century to determine the planet's rotation. He also postulated the inverse square principle that Isaac Newton applied to the law of gravity. When Newton's law was published, Hooke complained that he had not been given sufficient credit, and he became involved in a bitter controversy. After the *Principia Mathematica* appeared in 1686, Hooke made his charge of plagiarism. He had sent a letter to Newton in 1679, predicting that gravity decreases as the square of the distance travelled. The astronomer Edmond Halley tried to mediate, but Newton retaliated by removing almost every mention of Hooke's name from the next edition.

The Great Fire destroyed much of London, and Hooke was among the men who were asked to submit plans for the rebuilding of the city. Although Wren's scheme was adopted, Hooke worked with Wren as surveyor, and some of the buildings, such as the Monument, were designed by him. By the time he died in 1703, he had received many honours, but never the recognition due to a genius of magnitude equal to that of his more acclaimed contemporaries.

jar, the level of the water rose only a short way by the time the flame was extinguished or the animal died. He concluded that only a relatively small part of the air is suitable for maintaining either combustion or life, since he knew that the water level would rise as the atmospheric pressure fell within the jar. He then had to know if the active ingredient of air was the same for both flame and animal, or if two different fractions supported different functions. To discover this, he put both an animal and a candle into the same jar. They both died more quickly than the flame and animal which had been introduced singly. This proved that both were dependent on the same constituent of air. Actually, Mayow was discovering oxygen although he did not know it. Unfortunately, his findings got hopelessly confused with a concept of the time which was as much a red herring as pneuma.

At the time that Mayow was carrying out his experiments, scientists believed that when something burned it lost a mysterious substance which they called phlogiston. This concept was developed by a German chemist and physician Georg Ernst Stahl and held up the progress of chemistry for almost a hundred years. This fallacious theory resulted from the fact that when a piece of wood is burned, it seems to be reduced to a small pile of light ash. Actually, if all the smoke, water vapour, and gases could be collected the ash would gain weight, not lose it, as the result of oxidation.

Mayow wondered about the active component of air that supported respiration and combustion. He astutely compared it to the part of saltpetre or nitre which was used in gunpowder, since this has a built-in power of violently supporting the burning of sulphur and charcoal. Mayow called this nitroaerial spirit, which was actually his name for oxygen. But the false trail of phlogiston prevented recognition of Mayow's discovery for a hundred years and so this important finding went unheralded.

Then, in 1756, Joseph Black, a Scottish chemist, proved that when carbonates are heated they lose weight, due to the liberation of a gas he called fixed air. This was carbon dioxide and was the other side of the respiratory equation. Although Black's discovery truly destroyed the theory of phlogiston, that concept was too firmly established to fall under the first assault.

Twenty years later, Joseph Priestley, an English minister, an amateur scientist, and a devotee of the phlogiston theory, discovered a gas which he made by heating oxide of mercury; it was the same gas that Mayow had discovered. He found that flames burned with extraordinary vigour in this gas, which he called dephlogisticated air. In March 1775, he showed that a mouse lived longer in a given volume of his dephlogisticated air than it lived in ordinary air. Priestley later discovered nitrogen, the other main component of air, which he called phlogisticated air.

It was Antoine Laurent Lavoisier, the French chemist, who, in 1777, confirmed the nature of oxygen, which he showed appeared in inspired air, while the other gas, carbon dioxide, appeared in expired air. Lavoisier called the life-supporting gas vital air and later gave it the name oxygen. In the second volume of his memoirs, he showed that respiration decomposed the inspired air and that in cramped spaces the oxygen decreases and the carbon dioxide increases with a diminution in volume but an increase in weight.

At last the basic processes of respiration were beginning to unfold. Lavoisier put an end to the untenable phlogiston theory and, although he was not a medical man, applied his discoveries to practical uses, advocating, for example, a

Joseph Priestley, a nonconformist minister from Yorkshire, was born in 1733. Orphaned at an early age, he suffered from ill health and had a hard childhood. After his ordination, he met Benjamin Franklin and the American humanist inspired his interest in science. While still practising as a clergyman, he studied chemistry and electricity. His observations of fermentation at the brewery near his chapel in Leeds stimulated his discoveries of a number of gases, including oxygen, which he called dephlogisticated air. Priestley was elected a fellow of the Royal Society at the age of thirty-three and at about this time he developed a passion for political theory and wrote about the progress and perfectibility of humankind.

Priestley left his ministry, became librarian to Lord Shelburne, and accompanied his patron on his travels on the continent. In France, Priestley met Lavoisier and demonstrated his discovery of oxygen. In 1779, he left the service of Lord Shelburne and returned to chapel life in Birmingham. Priestley continued his scientific work and was the prime mover of the Lunar Society, a group of distinguished men who met each month to discuss science. So that they could travel more easily to the evening meetings, they always met at the full moon. The suspicious public referred to them as the "lunatics." Priestley began to write extensive treatises on theology and religion. In his *History of the Corruption of Christianity*, he rejected most of the fundamental doctrines of the faith. As with his political tracts, these writings aroused great public outrage. His sympathies were with the French Revolution and when he organized a dinner at his home in 1791 to celebrate the fall of the Bastille, an angry mob attacked his house, laboratory, and chapel, set them on fire, and destroyed everything.

Priestley emigrated to the United States. He was invited to be professor of chemistry at the University of Pennsylvania, recently founded by Franklin, but he refused. He became one of the early activists in the Unitarian movement and lived to the age of seventy-one.

minimum requirement of space for every person to inhabit, so that gas exchange could be adequate.

He explained respiration concisely: "We can state in general that respiration is but a slow combustion of carbon and hydrogen, similar in all points to that taking place in a lamp or a burning candle and that from this point of view animals which breathe are really combustible bodies which burn and are consumed. In respiration as in combustion it is the atmospheric air which supplies the oxygen, but in respiration it is the substance of the animal itself that provides the combustible material."

This was a vindication in a literal sense of the ancient concept of the *flamma vitalis*, the vital flame which burned in the heart. Lavoisier made one error; he assumed that the process of respiration, that is, heat-producing oxidation, took place in the lungs themselves.

It was Lazzaro Spallanzani, a professor of natural history at Pavia, who recognized that respiration occurs all over the body, in every tissue. He pointed out that the lungs are merely the organs of gas exchange, but they play no special role in the utilization of oxygen or the generation of carbon dioxide, which is a function of all the cells of the body. The blood is the circulating medium which transports the gases along with other substances.

In 1791, it was shown that the blood contains both oxygen and carbon dioxide. Then, in a work published in 1803, four years after his death, Spallanzani gave evidence that oxygen is absorbed and carbon dioxide produced throughout the body, and that this is the basic process of life itself, since the necessary energy is liberated in the conversion. Almost thirty years later it was proved quantitatively that the amount of oxygen and carbon dioxide is different in arterial and venous blood and that the change occurs in the tissues.

From this point onwards, detailed discoveries proceeded rapidly. It was found in 1856 that isolated muscles respire in the modern sense of the word, that is, take up oxygen and give off heat, producing energy for contraction by a process of oxidation. The next year the discovery was made that oxygen is held in some kind of chemical union and this was advanced five years later when it was found that the haemoglobin of the blood is the oxygen carrier.

From this latter third of the nineteenth century, discoveries came even more quickly. With increasing sophistication of laboratory equipment and knowledge of organic chemistry, the understanding of respiration, both in the sense of breathing and on the level of tissues and cells, accelerated on a logarithmic scale, laying the foundations for today's routine estimations for clinical purposes of concentrations of gases in the blood.

Antoine Lavoisier, the Father of Chemistry, provided physiologists with the key to respiration. Born in Paris in 1743, he was educated in the natural sciences and chemistry. His father, a lawyer, bought him a title in the Blois nobility. At the age of twenty-three, he won the gold medal of the Royal Academy of Sciences for his plans for the lighting of a large town and two years later he was elected a member of the Academy. Lavoisier married a girl of fourteen who became his assistant, wrote up his experiments, and entertained his large circle of friends. A portrait of the famous couple, posed amid scientific apparatus, was painted by David. Joseph Priestley's research interested Lavoisier and when the Englishman came to Paris in 1774 they met. Lavoisier grasped the meaning of Priestley's discovery; within three years he read his own paper on the composition of air to the Royal Academy. Later, as Commissioner of Powder, he did original research on explosives. Always active in public affairs, Lavoisier was a great philanthropist. As secretary of the government committee on agriculture, he designed a model farm in 1778 to demonstrate the importance of scientific farming methods. At the same time, he got involved in government financial policies, became a full member of the revenue-collecting agency, the Ferme Générale, and initiated reforms in taxation. During a famine, he provided a supply of grain to the province. Despite his innumerable contributions—scientific, humanitarian, and monetary—to France, Lavoisier's title and his commitment to the Ancien Régime earned him the distrust of Marat, and he was tried and sentenced to death by the revolutionaries. After five months in prison, he was taken to the Place de la Révolution and there guillotined on May 8th, 1794. His body was thrown into a common grave. When he heard of Lavoisier's death, Lagrange, the mathematician, said, "It only took a moment to sever that head, but perhaps a century will not be enough to produce another like it."

Lazzaro Spallanzani was born in 1729 near Modena, Italy, and received his early education with the Jesuits. Ordained as a priest, he then went to study law at Bologna. There he met a kinswoman, a famous professor of physics, who encouraged him to study the natural sciences. He became professor of logic and metaphysics at Reggio, and later professor of natural history at Pavia, where he was also curator of the museums. He died at the age of seventy. This outstanding polymath, whose great work was on digestion, was the first to experiment with contraception and artificial insemination in animals.

The Digestive System

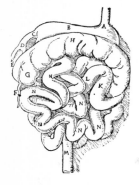

This drawing of the intestines, from Dryander's 1541 illustrated edition of Mondino's *Anathomia*, includes the earliest known depiction of the appendix, marked F.

Early interest in the gastro-intestinal tract was desultory and half-hearted. It aroused none of the philosophical passions that surrounded the elements, the humours, and the pneuma. Even to the most primitive and untutored observer, the body's digestive functions would seem to be among the most apparent. Food is put in at one end and food remnants, often easily recognizable, emerge from the other. Hunger, eating, and bowel action are so obviously related to each other that it is rather strange that in the history of anatomy and physiology the alimentary canal and its associated organs were relatively neglected until recent centuries.

The one exception was the liver, an organ which has always seemed to attract attention. But although the soothsayer used the liver, and sometimes the entrails, of animals for divination, including medical diagnosis, and the Babylonian clay and the Etruscan bronze models had recognizable contours corresponding to the general features of the liver, this did little for anatomy

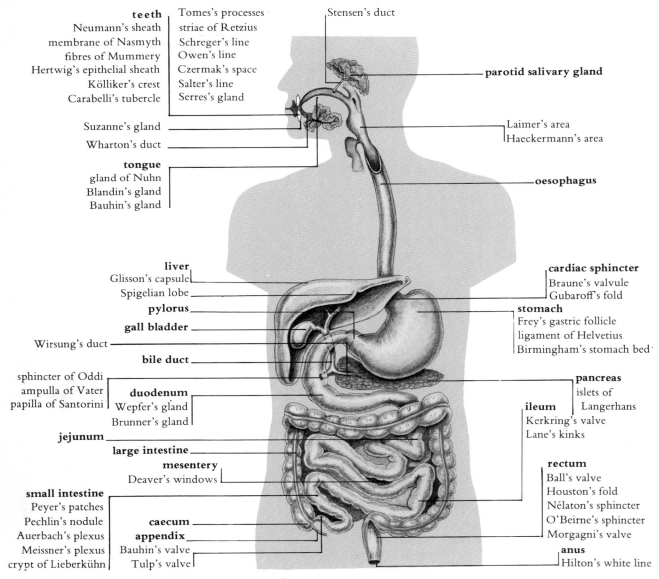

teeth
Neumann's sheath
membrane of Nasmyth
fibres of Mummery
Hertwig's epithelial sheath
Kölliker's crest
Carabelli's tubercle

Tomes's processes
striae of Retzius
Schreger's line
Owen's line
Czermak's space
Salter's line
Serres's gland

Stensen's duct

parotid salivary gland

Suzanne's gland
Wharton's duct

Laimer's area
Haeckermann's area

tongue
gland of Nuhn
Blandin's gland
Bauhin's gland

oesophagus

liver
Glisson's capsule
Spigelian lobe
pylorus
gall bladder
Wirsung's duct
bile duct
sphincter of Oddi
ampulla of Vater
papilla of Santorini
duodenum
Wepfer's gland
Brunner's gland
jejunum
large intestine
mesentery
Deaver's windows
small intestine
Peyer's patches
Pechlin's nodule
Auerbach's plexus
Meissner's plexus
crypt of Lieberkühn
caecum
appendix
Bauhin's valve
Tulp's valve

cardiac sphincter
Braune's valvule
Gubaroff's fold
stomach
Frey's gastric follicle
ligament of Helvetius
Birmingham's stomach bed
pancreas
islets of Langerhans
ileum
Kerkring's valve
Lane's kinks
rectum
Ball's valve
Houston's fold
Nélaton's sphincter
O'Beirne's sphincter
Morgagni's valve
anus
Hilton's white line

and nothing for physiology. No doubt butchers and hunters knew more about the gastro-intestinal system than did early medical men.

The physician Menon, a disciple of Aristotle, was the first to carry out experiments on nutrition and digestion. He wrote a book on medicine, which has long been lost, but lectures at the medical school in Alexandria were based on it and some fragments of a student's papyrus notebook, dated to about A.D. 150, have been found. These mention Menon's attempts to measure the gain and loss of weight of an animal in relation to its food intake. This is probably the earliest evidence of the application of quantitative methods in medical science.

In Alexandria, two notable men contributed a little to the knowledge of the gastro-intestinal tract. In the third century B.C., Herophilus of Chalcedon described, for the first time, the duodenum, the part of the small intestine which leads out of the stomach. Duodenum literally means twelve fingerbreadths, a reasonable estimate of its length. Herophilus discovered the lacteals, a word derived from milk, because these are the channels which carry the absorbed food from the intestine. After a fatty meal, the fluid, which is called chyle, is creamy white, a fact further elaborated by Herophilus's contemporary and rival Erasistratus, who observed the lacteals in newborn goats.

Herophilus also made some attempt at describing the anatomy of the liver, which a later Alexandrian, Rufus of Ephesus, also pursued. Rufus was responsible for the myth, however, that the liver has five, rather than two, lobes. His error was undoubtedly due to his assumption that the human liver is the same shape as the dog's liver. By A.D. 50, human dissection had probably ceased in Egypt and dissection would have been on lower animals. Rufus's mistake survived for one and a half thousand years; almost all later anatomists perpetuated the myth of the multi-lobed liver. Even in the sixteenth century Vesalius represented it so at first.

Galen, too, added to misunderstandings about the digestive system. In his extraordinary scheme of the bodily functions, the starting point was the alimentary canal. But he dealt with this in a perfunctory manner, so concerned was he with all the ebbing and flowing that he claimed went on from the liver and heart.

According to Galen, the nourishment absorbed from the food in the intestine came to the liver in the form of the milky liquid, chyle, that Erasistratus had described. This is largely true, but the chyle travels in the lacteals, part of the general lymphatic system. Galen believed that the chyle was carried to the liver by the portal vein, which actually contains blood. According to the Galenic system, the chyle was converted to blood in the liver, which imbued it with natural spirit. This misleading view held sway

for centuries and virtually no further progress was made in understanding the digestive system until the rebirth of observational anatomy and physiology in the late Middle Ages.

Almost nothing is known of the chronology and identity of the early describers of the individual parts of the digestive system until the era of the Italian, French, and German anatomists in the fourteenth to sixteenth centuries. Medieval illustrations survive which show crude dissection scenes, with barely recognizable organs strewn around. The liver is invariably a multi-lobed object and the intestines usually a shapeless mass. It must be remembered, however, that the material with which the early dissectors had to work was often putrified before they began. The abdominal contents would be the worst affected, a situation hardly likely to encourage prolonged exploration or detailed accuracy.

In 1326, Mondino, the Italian anatomist, published *Anathomia*, a handbook of anatomy and the first textbook which was entirely devoted to the subject. In it were his descriptions of the stomach, liver, and intestines, but, curiously, he omitted any mention of the appendix, which is so prominently attached to the caecum of the large intestine. He mentioned, however, a much less obvious structure, the duct of the pancreatic gland. But today this duct is usually known as Wirsung's duct, in commemoration of the man who described it three hundred years later.

Similarly, the duct of the salivary gland under the tongue is usually called Wharton's duct, but it was in fact first described by Alessandro Achillini in about 1500, half a century earlier than Wharton. Besides describing the salivary duct, Achillini made better recordings than Mondino and other early investigators of the caecum, duodenum, ileum, and colon.

This illustration, from a late thirteenth-century medical manuscript, is perhaps the earliest known portrayal of the actual process of dissection. The body of a woman has been opened and the organs removed so that the line of the vertebral column can be seen. The dissector, holding the knife in one hand and the liver in the other, is reproved by the physician and the monk. This picture, made at a time when dissection was just beginning to be practised, conveys the disapproval of the procedure which was still felt in some quarters.

It was not until Vesalius's *De Fabrica* appeared in 1543 that the digestive system was presented with any semblance of accuracy. The fifth book of the work was devoted to the abdominal contents and the standard of illustrations in both the main book and in the smaller *Epitome* are superb. Modern commentators on Vesalius have noted, however, that good though this section is, compared to the crudity of other, earlier writers, Vesalius passed more cursorily over these organs than might have been expected.

Although the appendix is illustrated no less than three times, there is not a word about it in the text. The book begins with a picture of the whole body with the abdominal wall removed, then each part—the stomach, omentum, intestines, liver, and gall bladder—are separately featured. Vesalius instructed that these smaller illustrations, all drawn to scale, should be cut out and pasted on the large figure of the body, to build up a kind of three-dimensional model, a technique still sometimes used today.

Many of the sixteenth-century anatomists published their books with illustrations, for better or worse, of the digestive organs. Johannes Dryander of Marburg brought out his book two years before Vesalius, yet it is obvious that some of the illustrations were plagiarized from the work done by the master at Padua. Dryander does show and name the appendix, calling it the additamentum.

Other writers of the same period included Charles Estienne (whose illustrations are mainly noted for their hideousness) and Spigelius, a Belgian and one of the last of the line of great anatomists at Padua. Spigelius's name is firmly attached to the liver. The caudate lobe, the knob-shaped lobule on the under surface, which the ancient soothsayers had noticed, is also known as the Spigelian lobe.

Spigelius described the caudate lobe in his

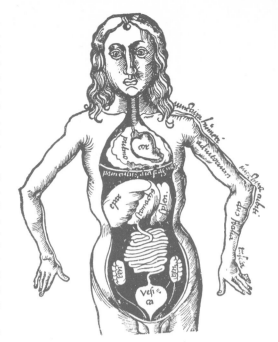

In the pre-Vesalian era, anatomical illustrations were crude and simplistic. Of indeterminate sex, this figure is from the *Margarita Philosophica* by Gregor Reisch, an illustrated anatomy text published in 1508.

own posthumously published book *De Corporis Humani Fabrica* of 1627, the same year as another eminent anatomist, Casserius, was posthumously published in Venice. Giulio Casserio, usually known as Casserius, succeeded Fabricius as professor at Padua and was one of William Harvey's teachers. His book contained excellent illustrations of the gastro-intestinal organs, which added considerably to the knowledge of the anatomy of the digestive system. He is best remembered for his work on the region of the ear, but the drawings of the abdominal organs are excellent. Rather than woodcuts, which were used by others up to that time, they were copperplate engravings.

The year 1627 was a vintage one for anatomy textbooks in Italy, especially posthumous books, for Gaspar Aselli's book *De Lactibus* appeared then. This was the result of his work on the lacteals. Aselli's text is notable for the large coloured plates, hand-tinted in every copy, which adorn it, the first time colour was ever used in a printed medical textbook. Opinion now has it that the exceptionally good pictures made up for the exceptionally bad text.

The anatomy of the gastro-intestinal tract and its associated organs and glands carry the names of many men who described the various parts.

A number of glands in the mouth and throat were examined closely in the seventeenth century. Although the prominent papillae under the tongue, where the ducts open, had been

Adriaan van der Spieghel, better known by the Latinized name Spigelius, was born in Brussels in 1567. He studied with Fabricius and Casserius at Padua and served as state physician in Moravia and as professor of anatomy of Venice before returning to Padua to take the chair of anatomy. Spigelius contributed to the important work of standardizing anatomical nomenclature, which was still a confusion of Arabic, Latin, and Greek names and he gave the first adequate description of the spinal muscles. His major work, *De Humani Corporis Fabrica*, was published in 1627, two years after his death.

noticed at a much earlier date, the English anatomist Thomas Wharton described the duct of the submandibular salivary gland in his book on glands, *Adenographia*, published in 1656.

Wharton, on the basis of his considerable work, declared that every gland in the body must have its secretory duct. He was right in regard to the exocrine glands, those that produce their secretions externally. But, of course, at that time there was no knowledge of endocrine glands, those that release their secretions directly into the bloodstream to act on distant organs and tissues.

Wharton's statement about ducts encouraged the Danish anatomist Niels Stensen to do a more detailed study of glands and it was he who located the parotid (from the Latin "beside the ear") salivary glands and the duct which now bears his name.

In the same century Anton Nuck, who is best remembered in the canal of Nuck in the female generative system, described smaller salivary glands in his book of 1685.

The glands of Nuhn, on the front of the tongue, which also produce saliva, are sometimes called Blandin's glands because Philippe Blandin, a surgeon in Paris, described them in 1823, while Anton Nuhn, a professor of anatomy in Heidelberg, did not publish his account until 1845. These same rather insignificant glands are also graced with other eponyms. As Bauhin's glands, they commemorate Caspar Bauhin who was a professor, not only of medicine and of anatomy at the University of Basel, but also of Greek and of botany, as well as being physician to the Duke of Wittenberg. He described the tongue glands in 1590. So did Tigri in 1847, Breda in 1886, and Zincone in 1877, and each of their names are occasionally used in reference to these glands.

A gland in the floor of the mouth rejoices in the name Suzanne's gland, but the name refers to a prosector of pathology at Bordeaux, Jean Georges Suzanne, who described it in 1877.

The area where the gullet, or oesophagus, joins the pharynx, the deepest part of the throat, has been named twice for different Germanic anatomists. It is sometimes called Haeckermann's area after Karl Haeckermann, an anatomist of Bremen who described it in 1890. But it may also be called Laimer's area after Edouard Laimer, a prosector of anatomy at Graz in Austria, who described it seven years earlier.

At the junction between the gullet and the stomach, a ring of muscle fibres acts as a valve. This is called the cardiac sphincter, because it is near the heart, just across the thickness of the diaphragm. A fold in the lining of the oesophagus over the cardiac sphincter is sometimes called Braune's valvule, after Christian Braune, a professor of surgery and anatomy at Leipzig, who described it in 1875. The same structure is

Gaspar Aselli held the chair of anatomy at Pavia. Born in Cremona in 1581, he spent his life in Italy. His great work was on the lacteals and his notable discovery of them was made during a public post-

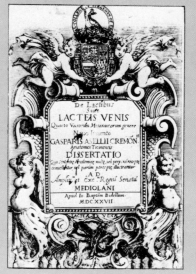

mortem on a dog in 1622. At first he thought the prominent white cords were nerves, but when he cut them milky fluid gushed out and he is reputed to have shouted "Eureka!" Aselli died in Milan in 1626, but his observation on this and subsequent experiments appeared as a book, *De Lactibus Sive Lacteis Venis*, published posthumously. His work was, however, repudiated by Harvey.

called Gubaroff's fold or valve because it was described a few years later by Alexander Petrovich von Gubaroff, a Russian professor of gynaecology in Moscow. There is no record to explain how a gynaecologist came to name a structure in the throat.

In the stomach itself, there are many thousands of tiny glands which produce hydrochloric acid and the enzyme pepsin. These glands open into the gastric lining of the stomach by tiny dimples which are called Frey's gastric follicles. They are named after Heinrich Frey who was born in Frankfurt in 1822, and became professor of histology at Zürich.

In the wall of the stomach, outside the follicles, are the bands of muscle fibres and fibrous tissue which give mobility and support to the organ. They are known as the ligaments of Helvetius because they were first described by Johannes Helvetius, a comparative anatomist who wrote extensively on the structure of animal stomachs. Helvetius was born in Paris in 1685 and became Royal Physician as well as State Counsellor to the Queen of France.

One of the most intriguing names in the body is concerned with the area around the stomach. Ambrose Birmingham, a Dublin Irishman who was a professor of anatomy at the Catholic University of Ireland, published a paper in 1896 concerning the topography of the upper part of the abdomen. In it he described the various surfaces on which the stomach rests as Birmingham's stomach bed.

The pylorus is the ring of muscle which keeps the exit from the stomach tightly closed until the contents are in a fit state to be allowed into the duodenum for further digestion. There are a number of small glands in the wall of the duodenum. These are known as either Brunner's glands or Wepfer's glands. Wepfer was, incidentally, Brunner's father-in-law. Johann Jacobus Wepfer was born in Schaffhausen, Switzerland. He studied medicine in nearby Basel, then in Strasburg. Wepfer, who became State Physician in Basel and doctor to some of the German princes, was a great believer in post-mortem examination and was the first man to recognize the relationship of cerebral hae-morrhage to apoplexy. His daughter married Johann Konrad Brunner, who was also a Swiss. Professor of anatomy first at Heidelberg and then at Strasburg, Brunner described the duode-nal glands in 1687, although his father-in-law had named them some years earlier.

The pancreas, the gland that lies horizontally beneath the stomach, was not investigated in any serious way until late in the evolution of anatomy and physiology. Its internal secretions —for it is both an exocrine and an endocrine gland—were unknown until the late nineteenth and early twentieth centuries. The external secretion of strong digestive juice was, however, discovered much earlier. The pancreatic duct that leads the secretion into the duodenum was

drawn, although not described, in 1642 by the German anatomist Johann Wirsung, whose name has remained attached to the duct ever since. Wirsung, a professor of anatomy at Padua, was assassinated in 1643, apparently as the result of a quarrel over priority in the dis-covery of the pancreatic duct.

In common with the bile duct, the duct of Wirsung emerges in the duodenum. They do not actually join, but run together at the last part of their course, sharing a common outlet on the tip of a little mound which projects into the duodenum. This outlet is called the ampulla of Vater. An ampulla was a small flask used in Roman times. Vater was Abraham Vater, pro-fessor of anatomy, botany, and, later, pathology at the University of Wittenberg in Germany. He described his ampulla in 1720.

The mound on which the ampulla opens is sometimes known as the papilla of Santorini. A papilla is a nipple, and Santorini was the eighteenth-century Venetian anatomist.

The mouth of the pancreatic and bile duct is closed by a circular ring of muscle fibres called the sphincter of Oddi. Although he was not the first to describe it, Ruggero Oddi, a doctor at the University of Perugia, is usually associated with the structure which he described in 1887.

The sphincter was first described by Francis Glisson, but he is best known because of the thick but tough covering of the liver, known as Glisson's capsule, which he named in 1654. Glisson was one of the great clinical physicians of the seventeenth century and was the founder of a physiological doctrine known as "irritabil-ity." Glisson believed that the human body was capable of being stimulated by some external influence, to which it then reacted appropriately. Glisson applied his theory to the liver, in which he was particularly interested. He concluded that the periodic emission of bile was due to irritation of the bile vessels. He projected this hypothesis to the life process itself. "Every part that suffers from an incommodity seeks to dis-embarrass itself," he wrote. His elaboration of his materialistic and mechanistic theory was typical of attempts of the time to devise models for the understanding of the body's vital processes.

The gall bladder has always been an organ of interest, witnessed by the prominence that bile, whether green or yellow, had in the theory of the humours. The bladder itself is a noticeable structure, bulging below the lower edge of the liver. It is not known who first described it, but gall is a later Germanic-derived name. The earlier root, *chol*, as used in choleric, cholesterol or cholecystectomy (surgical removal of the gall bladder), comes from the Greek, *khol*, meaning bile, the thick green liquid that was so obviously one of the main features of the liver's manufac-turing activities.

Francis Glisson was one of the great seventeenth-century English physicians. Born in Dorset in 1597, he went to Cambridge where he read classics, then studied medicine under Harvey. At the age of thirty-nine, he became Regius Professor of Physik at Cambridge, one of the eminent posts he filled during his long life, for he lived to the age of eighty. In 1650, Glisson published his important paper on rickets. Four years later, he published his classical account of the liver. His later publications on general physiology and irritability were a sign of the emergence of the scientific search for wider truths. When Harvey retired as Lumleian Lecturer at the Royal College of Physicians, Glisson, who was one of the founders of the Royal Society, took up the post. During the Great Plague of 1665, he was one of the few doctors to stay behind in London.

In 1667, his achievements were further recognized when he was elected president of the Royal Society. Influenced by his studies in philosophy, and by Aristotle in particular, he tried to formulate a unifying theory to explain the development of all physical matter out of the fundamental energy of nature.

In the lining of the intestine there are folds known as Kerkring's valves. Though Fallopius described them long before, the full name valvulae conniventes was given by a German anatomist Theodor Kerkring in 1670.

Another so-called valve lies between the small and large intestine, where the ileum meets the caecum. This ileo-caecal valve is known either as Bauhin's valve or Tulp's valve. The first was Caspar Bauhin who also described the gland on the tongue. But the valve is also named after Nicholas Tulp, a contemporary of Bauhin, from the Netherlands, who described it in 1641.

Some curious eponyms exist in relation to the intestine. These include Lane's kinks, Deaver's windows, and Peyer's patches. Lane's kinks refer to inconstant twists in the intestines, up to six in number, the most prominent of which is in the ileum. In the *British Medical Journal* of 1911, these twists were described in an almost satirically titled article, "The Kinks Which Develop in Our Drainage System in Chronic Stasis," by the eminent senior surgeon at Guy's Hospital, London, Sir William Arbuthnot Lane.

Deaver's windows are translucent areas in the mesentery, the membrane that fixes the intestinal canal to the back of the abdominal cavity and carries the blood vessels to and from the gut. Mostly thickened with fat, spaces appear here and there which fancifully resemble windows. They were so named in 1901 by John Blair Deaver, an anatomist and surgeon from Philadelphia.

Peyer's patches are the islands of lymphatic tissue that occur at frequent intervals in the lining of the small intestine and are the focus of attack during typhoid fever. They were named after Johann Conrad Peyer, a German anatomist. Born at Schaffhausen on the upper Rhine in 1653, Peyer studied in Paris and returned to Schaffhausen where he became professor of logic, rhetoric, and medicine. He also described smaller nodules of lymphatic tissue in the intestine, but another Johann, a Dutchman called Pechlin who was professor of medicine in Kiel and Stockholm, named them earlier, in 1672.

The snake-like contractions of the intestine, called peristalsis, from the Greek "to send around," move the food contents along by a co-ordinated wave-like squeezing of the muscle coats of the gut itself. These movements are under the control of the autonomic nervous system which cannot be influenced at will. The nervous activation is brought about by two gossamer-fine networks of nerve fibres buried between the layers of the intestinal wall. The first is deep down between the lining and the inner muscle coat, the other lies between the muscle layers. They are called Auerbach's plexus and Meissner's plexus.

Leopold Auerbach, a German professor of neuropathology, described what in modern

Jean Baptiste van Helmont was born in Brussels in 1577. A physician and lecturer in medicine and chemistry at the University of Louvain, he was also a Capuchin friar. Caught between systematic investigation and religion he was the Father of Biochemistry and he had mystical visions. Although he was a follower of the iatro-chemical school, he made a number of important observations. He devised one of the early thermometers and investigated the use of the pendulum for the measurement of time. He invented the word "gas" and noted that acid is concerned with digestion in the stomach.

terminology is called the myenteric plexus, which lies outermost. He described his work on the innervation of the gut in 1862, almost a decade after Georg Meissner, a German professor of physiology, published on the submucous plexus.

It is strange and somewhat ironic that the eponyms of so many structures are not the names of the men who originally described them. In the case of one feature of the intestines, it was no less than the fourth man in line who acquired immortality. Along the length of the small intestine are tiny pits in the lining which are variously called glands, crypts, or follicles. The most common eponym for them is the crypts of Lieberkühn. It was the great Malpighi, however, who first described them in 1688, then Brunner in 1715, and later Galeati in 1731. Yet Johann Nathaniel Lieberkühn, who did not mention them until 1785, has his name commemorated there. Lieberkühn was a German physician and anatomist who was born, lived, and died in Berlin between 1711 and 1756.

It is at the terminal part of the alimentary canal that surgeons, rather than anatomists, have created an eponymy out of proportion to the size of the area. Ball's valves, for example, are folds in the lining of the rectum. Previously described by Morgagni, they are now attributed to Charles Bent Ball. Born in Dublin in 1851, he became Regius Professor of Surgery there and a member of the General Medical Council. He was Surgeon to the King in Ireland, was knighted in 1903, and elevated to the baronetcy in 1913.

The rectum abounds in valves and folds. Another set is Houston's folds, after another Dublin surgeon, John Houston. Then there is Nélaton's sphincter which commemorates Auguste Nélaton, a Parisian professor of surgery in the mid-nineteenth century who became sur-

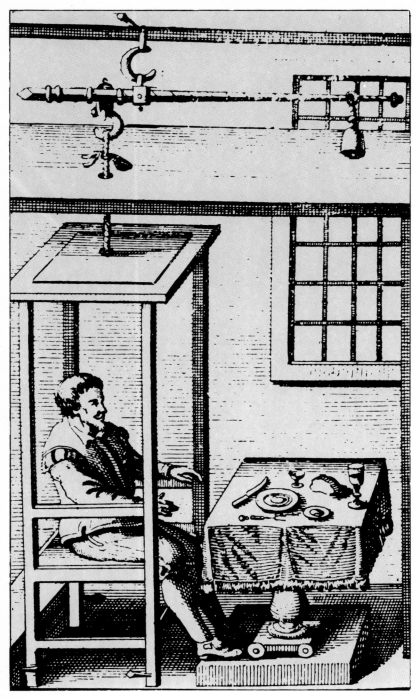

Sanctorius was a pioneer of measurement in medicine and was the first to investigate metabolism. For thirty years he worked, ate, and slept in this contraption, which he designed and constructed, noting variations in his body weight, relative to his food and fluid intake.

named after John Hilton, an illustrious London surgeon. A lecturer and surgeon at Guy's Hospital, he became one of the original fellows of the Royal College of Surgeons in 1843 and its president in 1867.

Quite apart from the anatomy of the gastro-intestinal tract, many medical investigators tried to make sense of the process of digestion. One of the most interesting was Santorio Santorio, who is better known by his latinized name Sanctorius. Born in the Italian town of Capodistria in 1561, he became a professor of medicine at Padua. Sanctorius was of an inventive and ingenious turn of mind and, in the same manner as Harvey, he applied quantitative measurements to physiological problems.

Sanctorius was a friend of Galileo, and from him he borrowed the idea of using the thermometer and pendulum in medical investigations. He developed the first clinical thermometer, which he said could be used to "learn daily with accuracy how the patient's heat deviates from normal." He also devised an instrument to measure the pulse rate, using a pendulum, the length of which could be adjusted to synchronize with the beat of the pulse which was indicated on a dial. This was a hundred years before any similar device was employed. Sanctorius was, indeed, a man before his time, and the world was not yet ready for many of his ideas.

In his investigations of the process of digestion, Sanctorius attempted to measure with accuracy the loss and gain of weight relative to food and fluid intake and to make allowances for activity and all other functions. To achieve this, he constructed special scales, in which a large wooden frame was suspended from a lever balance in the ceiling. Into this frame he was able to put his table, desk, bed, or other pieces of furniture. His usual device was a suspended chair set before a fixed table.

For thirty years Sanctorius took careful measurements of the food and drink he consumed and the weight of his waste products. There was, naturally, a difference between input and output. This he ascribed to "insensible perspiration," by which he meant not only fluid loss due to sweating, but mass loss due to the conversion of food and body tissue into energy. He never really understood the complexities of what was happening within his body, but this was nevertheless a brave and imaginative attempt to bring measurement into physiological experiments. Indefatigably, in sickness and in health, before and after sleep, meals, exercise, and even sexual intercourse, he weighed himself in his apparatus.

In 1614, after thirty years of investigation, the results of Sanctorius's findings were published in a book of aphorisms called *De Statistica Medicina*. It became a best-seller of its day and

geon to Napoleon III. The same structure carries the alternative name of O'Beirne's sphincter for James O'Beirne, another Dublin surgeon who, like Ball, was Surgeon to the King in Ireland earlier in the nineteenth century. O'Beirne described the sphincter in 1833 in a paper entitled "New Views of the Process of Defaecation."

Morgagni also has his name attached to valves in the anal canal. A more curiously named structure there is Hilton's white line, a demarcation zone of significance to surgeons which was

was reprinted in many editions and translated into many European languages.

In addition to his pioneering methods of investigation, Sanctorius was the major proponent of one of the odd medical cults that sprang up in the seventeenth century and were symptomatic of the realization that there was a need to replace the recently discredited Galenism with something new. Just as ancient Rome had had schools of empiricists, pneumatists, and methodists, the Renaissance had its schools of often eccentric scientific philosophies. One of these was led by Francis Glisson with his doctrine of irritability. Sanctorius was the leader of the iatrophysical school, which tried to explain all human function in terms of a machine. His balance experiments were an attempt to reduce human metabolic processes to a quantitative system.

Although Sanctorius's work and that of the various other schools were themselves useless in advancing knowledge, they acted as a bridge from the stagnant theorizing of the past to modern scientific methodology, and, as such, invite applause in spite of their impracticality.

Another member of the iatrophysical, or iatromechanical, school, was Gian Alfonso Borelli who, like Sanctorius, believed that the body could be reduced to a mechanical analogue. Borelli became professor of mathematics at Padua, but had an interest in things biological, as did most polymaths of that time. Borelli was among the iatromechanists who believed, much as many modern scientists do, that the body is a machine, but the soul is the driving force and the cause of all life processes.

Borelli was more interested in muscular movements than Sanctorius, who was more concerned with metabolism. Borelli explained digestion as a process of "trituration" in the stomach, by which he meant a grinding action which reduced food to particles small enough to be absorbed. This theory was in opposition to that of the school of iatrochemistry, who used the words effervescence, fermentation, and even putrefaction, to describe the means by which food was rendered suitable for absorption.

The leader of the iatrochemists was Jean Baptiste van Helmont, a Capuchin friar from Brussels. He was a pioneer in the study of the chemistry of gases and invented the very name gas, which he derived from the Greek word for chaos. An odd character, he refused an M.A. degree on one occasion saying that he was not a master of any arts. He also refused to devote his life to theology, because he maintained that, if he did, his livelihood would depend on the sins of the people.

Another iatrochemist, the founder of the school, was Sylvius, a Dutch scientist whose real name was François de la Bois. Franciscus Sylvius is prominently remembered in the anatomy of the brain, but he also made a great contribution to the understanding of the digestive processes, although he did try to graft contemporary thought on the old Galenic framework. Sylvius stressed the importance in digestion of the triad of saliva, pancreatic juice, and bile. This is now known to be true, but his ideas about effervescence and fermentation were wide of the mark.

Regnier de Graaf, the Dutch anatomist whose name is principally remembered in connection with the Graafian follicles of the ovary, made a number of experiments on pancreatic juice in the seventeenth century. He made a fistula, an artificial opening, from the pancreatic duct of Wirsung on to the skin of the abdominal wall of a dog, so that he could collect the digestive fluid as it was produced. His description of this experiment was published in 1664. Among his many other original experiments, de Graaf collected bile from a biliary fistula and studied the effects of saliva on foodstuffs. Saliva and bile contain substances which are important in the digestive process.

René Réamur was another investigator of the digestive system. He, too, invented a clinical thermometer. A Frenchman, it was said that "He was one of the most striking men of science of the eighteenth century and, indeed, in some respects, of all time." Réamur's scientific abilities covered many fields, including mathematics, chemistry, and physics. His famous thermometer and Réamur scale, which is still used, were produced in 1731. But he was also a biologist and in 1752 he wrote an outstanding work on *Digestion in Birds*. By clever experiments on various animals, he was able to retrieve food at different stages of digestion and study the progress of the digestive process. For much of his work he used a large bird, the kite, since it could be easily induced to regurgitate its food

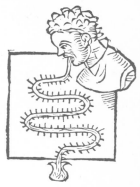

Sanctorius was inspired by Galileo to devise the first clinical thermometer to record body temperature. The patient was required to hold one globe-shaped end in his mouth, while the serpentine coil of the thermometer, graduated with glass beads, led down to a bulbous end immersed in a bowl of water.

François de la Bois, known as Sylvius, was one of the leading anatomists at the University of Leyden, the university that rivalled Padua in the seventeenth century. Born in Frankfurt in 1614, of a French family, Sylvius lived, studied, and taught in northern Europe until his death in 1672. He qualified in medicine at Basel, settled in Amsterdam, and, in 1658, was made professor of practical medicine at Leyden. An outstanding teacher, he said: "I have put the symptoms of disease before their eyes, have let them hear the complaints of the patients, and asked their opinion in every case." Sylvius was a Galenist, but because he was also an advocate of the new scientific method he supported William Harvey.

Discovering the Human Body

when required. Réamur was able to disprove the theory of putrefaction as part of digestion.

Lazzaro Spallanzani carried out similar experiments but used himself as the subject instead of a kite. An Italian, Spallanzani had wide-ranging interests, especially in embryology and regeneration of tissues. His work on fertilization in animals was profound and he was well known for his work and his writings in opposition to the popular theory of spontaneous generation, the concept that living creatures would appear from non-living material, like maggots from rotten meat or worms from compost heaps. By numerous experiments, Spallanzani showed that some type of egg or seed must always be present. Spallanzani's great work, however, was on digestion, published in a dissertation in 1780.

He interpreted the process of digestion by retrieving partly digested food from his own stomach and by tying threads to morsels and making himself vomit. He also swallowed porous bags and tubes, which he later recovered from his stools. Spallanzani found that, though mastication in saliva and Borelli's trituration in the stomach were useful preliminaries, true digestion required dissolving in the stomach and intestines. He came very near rediscovering the acid nature of the stomach contents, which had been forgotten since the time of van Helmont.

In 1803, John Richardson Young, an American, wrote a thesis for the University of Pennsylvania on the process of digestion. It was he who showed that acid gastric juice had a dissolving effect on most foods. He proved that Borelli's accepted theory of trituration was incorrect and that trituration could more accurately be applied to the process in a bird's crop, where hard cereal grains and seeds are milled with the aid of grit particles in the strong muscular organ. In mammals, however, it is a solvent action that is more important. Young explained that acid is essential to the process of digestion. (The identity of the acid as hydrochloric was made by William Prout in 1824.)

Another American, William Beaumont, made an important contribution to the investigation of the digestion. Although he had never been to medical school, Beaumont was a surgeon in the United States Army. In June 1822, Beaumont was called to the trading post on Mackinaw Island where Alexis St. Martin, a young Canadian traveller, had received an accidental gunshot wound in the abdomen.

When Beaumont first saw the young man, he thought that death was inevitable. He did his best to replace parts of the lung that were bulging through the gaping wound and he pushed the abdominal organs back into place, putting a compress over a hole in the stomach large enough to admit the doctor's finger. Beaumont considered this merely a palliative treatment to make the young man's last days more comfortable.

But good health and the strong physique of Alexis St. Martin pulled him through. It was not too long before he was perfectly well, except for an ugly scar across his chest and abdomen and a permanent fistula into his stomach. The fistula was open to the outside so that he had to cover the hole during and after meals to stop food from escaping.

Beaumont became intrigued with his patient's unique defect and for some years took advantage of it to study the process of digestion. He asked Alexis to swallow different foods, such as raw and cooked meat, tying pieces of silk thread to them so that they could be pulled up again for examination. He collected the stomach contents and digestive juices by means of a tube passed into the fistula.

In 1833, after seven years' work, the length of

time due in part to Alexis's frequent trips to Canada, Beaumont was able to publish, at his own expense, a book called *Observations on the Gastric Juice and the Physiology of Digestion*. In the book Beaumont described the movements of the stomach; he confirmed the presence of hydrochloric acid and of a ferment shown thirteen years later to be the protein-breaking enzyme, pepsin. His fifty-one conclusions are valid today, making William Beaumont one of the founders of modern metabolic physiology.

Beaumont influenced many other investigators. One of these men, Claude Bernard, the great French scientist, was the pioneer of liver physiology. The functions of the liver were little understood until recent times. Although it was given an important place by the ancients as the seat of blood formation, little further attention was offered until methods of microscopy and biochemistry became available. By 1857, however, Bernard had shown that the liver in an animal produces sugar and glycogen, no matter what type of food is ingested. His work was extended by his pupils, including Willy Kühne, the founder of the science of enzymes.

Bernard also used animal fistulae to examine the process of digestion. He made surgical fistulae on many dogs. Until the results of his work were published, it was thought that all digestion took place in the stomach. But Bernard showed that only the preliminary stages of digestion occur in the stomach, in an acid medium. The rest take place in the intestines after the intensely alkaline bile has neutralized the stomach acid.

While engaged in this work, one of Claude Bernard's experimental dogs escaped from his laboratory with a tube still fixed to the hole in its belly. It ran home and was brought back by its angry owner, who happened to be an inspector of police. The dog must have been press-ganged into becoming a laboratory subject, and Bernard managed to placate the policemen, closed the fistula, and returned the dog to its owner without ill effects.

After Bernard, other investigations were performed with stomach fistulae, and these investigations became so well known, even to the lay public, that the name Pavlov is now almost a household word.

Ivan Petrovich Pavlov, a Nobel Prize winner in 1904, was a Russian experimental surgeon who investigated the nervous control of digestion by making fistulae into the stomachs and pancreatic ducts of dogs. He was thus able to collect the gastric and pancreatic juices under a wide variety of conditions. His most memorable discovery is that the production of digestive juices can be stimulated without actual food being introduced into the stomach. This had always been known in the case of the salivary glands—the mouth-watering at the sight or even

Claude Bernard, who continually demonstrated the value of the experimental method, was gifted with a persistent and creative genius which led him to make discoveries in many fields of medicine. Born in 1813, in Villefranche, he was educated by the Jesuits. His first job was as a pharmacist's assistant. He then tried his hand as a playwright, but a critic persuaded him to take up medicine. Bernard studied under François Magendie, whom he succeeded as professor of medicine. He was also appointed to the new chair of physiology at the Sorbonne. When he died, in 1878, Bernard was given the first public funeral ever accorded to a man of science in France.

the thought of food. Pavlov showed that similar stimulation occurred in the other less accessible digestive organs and that these secretions are under nervous and psychic control, triggered by stimuli under such predetermined circumstances as the ringing of a bell after a dog had been appropriately conditioned.

The story, then, of the discovery of the anatomy of the digestive system is one of piecemeal and sporadic advance, often strongly influenced by the practical needs of surgeons, whose happy hunting ground has traditionally been the abdominal cavity, in which most of the gastrointestinal system lies. The discoveries of function came comparatively late and proper understanding of it, especially in its biochemical complexities, has only been achieved in the late nineteenth and twentieth centuries.

Ivan Petrovich Pavlov is one of the few physiologists whose name is a household word. He was born in central Russia in 1849. He intended to enter the priesthood, but in 1870 he left the seminary and went to the university in St. Petersburg, where he studied chemistry and physiology, then to Leipzig, where he worked under Carl Ludwig and Rudolf Heidenhain. Pavlov held the chair of surgery at the Imperial Medical Academy and in 1904 was awarded the Nobel Prize in medicine for his research in the physiology of digestion. He continued to work in Leningrad until his death in 1936.

Teeth

The first accurate dental diagram was probably this one which appeared in the first volume of Vesalius's *De Fabrica*. The molar on the left is laid open to expose the pulp cavity, thus disproving Galen's claim that teeth are solid.

Considering that teeth are the first mechanism in the process of digestion, it is rather surprising that the practice of dentistry as a respectable profession is relatively new. Until recent centuries, the drastic ministration of itinerant tooth-pullers offered the only remedy for most dental ailments. True, the Egyptians seem to have filled teeth with molten lead, and both the Etruscans and the Romans used primitive sets of false teeth. Celsus, writing in the first century A.D., described the wiring of teeth and the use of a dental mirror. But, apart from a few such references, little is known of early dental practice.

Aristotle seems to have been the first to record the structure of teeth. Curiously, he mixed penetrating accuracy of observation with gross errors. He was perceptive enough to note that animals with horns never have teeth in the front of the mouth, and yet he seriously asserted that women have fewer teeth than men.

Galen also wrote on the structure of teeth, although he seems to have thought there are only sixteen — an extraordinary assumption when all he had to do was count. This error was quoted by the fourteenth-century surgeon

Guy de Chauliac, who alleged that there also were "often" thirty-two teeth, although there were "often" only twenty-eight. This peculiar statement seems typical of the medieval mind; even the simple counting of teeth was subjected to wild, theoretical speculation.

Early errors as to the number and form of teeth were reproduced faithfully by generations of anatomists, none of whom seemed aware of the possibility of unerupted wisdom teeth and many of whom stated that the permanent teeth grew from the roots of the milk or deciduous teeth. Even in the sixteenth century, Vesalius perpetuated this fallacy, for which he was roundly criticized by his contemporary Eustachius, who revealed the truth — that the nuclei of the permanent teeth lie in cavities in the jaw, quite unconnected with the milk teeth.

In *De Fabrica*, Vesalius recorded that Galen had believed teeth to be different from bone, because, unlike bone, they had the property of sensation and pain. Vesalius was bold enough to contradict the Galenic view that there are only sixteen teeth and pointed out that anyone could disprove it by counting his or her own. The book includes an illustration of a molar

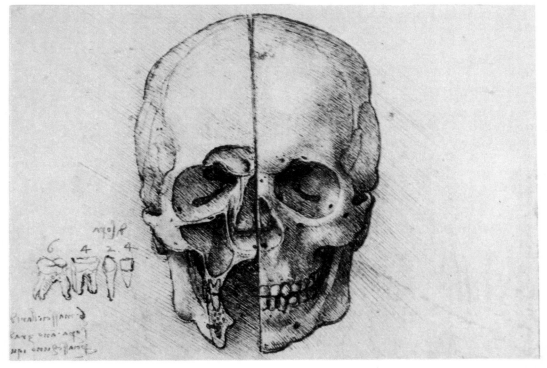

Unlike early anatomists, Leonardo, the artist with a scrupulous concern for fine anatomical detail, does seem to have counted teeth. Shown alongside his bissected skull, which has a full complement of teeth on one side, are four numbered molars and incisors, drawn separately to illustrate the roots.

tooth in cross-section, showing the central pulp cavity, probably the first true dental diagram in existence.

The most significant early text on dental anatomy was Eustachius's *Libellus de Dentibus*, published in 1563. In this he described foetal and infant dentition and the minute structure of teeth, with an outer marble-like layer on top of a less compact core. This is the first description of the enamel and the dentine. There was much speculation as to how the teeth were nourished. Eustachius wasn't sure, but Vesalius correctly deduced that it was from the pulp cavity. The nervous sensitivity of the teeth—such a marked feature in toothache—was also a source of endless speculation. Fallopius included some dental observations in his book of 1562, including a description of the tooth socket.

The word enamel was first used for the outer shell of the teeth in an English book *The Operator for the Teeth*, printed in York in 1685. The author, Charles Allen, gave a good description of the minute structure of the teeth from studies he had made using a simple lens. Yet, seven years earlier, Anton van Leeuwenhoek, who developed the microscope, had written to the Royal Society about the channels and pores in the substance of teeth. This important communication was neglected and forgotten for almost a century. In 1771, John Hunter, the celebrated English anatomist and surgeon, published *The Natural History of the Human Teeth*, and this was followed thirty-two years later by another book with exactly the same title written by Joseph Fox.

From the early nineteenth century onwards, it was the microscope, together with other technological aids, which advanced the knowledge of teeth—and raised the practice of dentistry from the level of fairground burlesque to a profession equal in status to medicine. More recently, the two professions have forged closer links, in that the speciality of surgery of the face and jaw often requires the expertise of both. In many European countries dentistry is not a profession apart, but is practised by stomatologists—qualified doctors who have specialized knowledge of the jaws and their contents.

A tooth consists of the flint-like enamel, the hardest and most durable of all the body tissues, the hard but porous dentine, and the soft pulp which is filled with blood vessels and nerves. Most of the eponyms in dental anatomy concern the inner structure of the tooth.

As with anatomical eponyms in general, named parts in the teeth sometimes commemorate more than one member of the same family. Tiny projections of the cells which produce enamel substance, for example, are called Tomes's processes, after Sir Charles

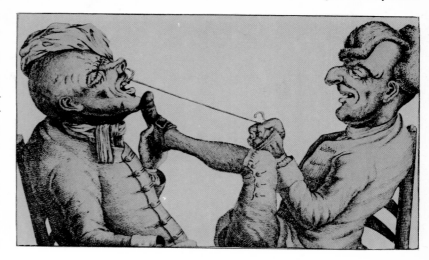

Sissmore Tomes. He edited a textbook of dental anatomy written by his father Sir John Tomes, whose own name is attached to fibrils, the tiny filaments that make up dentine. Both father and son were on the staff of the London Dental Hospital and each lived to be more than eighty years old.

Gustave Magnus Retzius succeeded his father, Anders Adolf Retzius, as professor of anatomy at the world-famous Karolinska Institute in Stockholm. Both men have a number of structures named after them, including the striae of Retzius, brown lines in the enamel, which commemorate Gustave. There are many eponyms in the dentine. Among these are Schreger's lines, which are identical with Owen's lines. Schreger was a German chemist. Sir Richard Owen was curator of the museum of the Royal College of Surgeons. Also in the dentine are Czermak's spaces, the fibres of Mummery, Neumann's sheath, and Salter's lines.

The enamel layer has fewer associated names. The cuticle of the enamel was described in 1839 by Alexander Nasmyth, who was dentist to Queen Victoria and Prince Albert. Besides being remembered for this membrane of Nasmyth, he was renowned for his original work on the surgical treatment of cleft palate. Hertwig's epithelial sheath, the remnant of the enamel-forming tissue which sometimes persists around the roots of teeth in infants, is named for Oscar Hertwig, a German professor of anatomy. The same groups of cells are called Serres's glands, commemorating Antoine Serres, an eighteenth-century Parisian anatomist better known for his theories of evolution.

Other eponymous features in the teeth include Kölliker's crest, the part of the upper jaw where the front teeth often arise, named for a professor of surgery at Leipzig early in the twentieth century; and Carabelli's tubercle, an inconstant extra cusp on one of the molars, which commemorates a Hungarian-born professor of dentistry in Vienna.

The popular view of the itinerant tooth-puller is epitomized in this lively cartoon published in Manchester in 1773. It was only in the early nineteenth century that dental practice began to achieve respectability.

The Urinary System

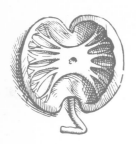

Vesalius illustrated the kidney of a dog in *De Fabrica*, because he felt the human kidney was too fatty to show the mythical sieve-like structure which Galen had postulated as the renal filtration unit. It was not until the discovery of the structure and function of the nephron centuries later that the sieve theory was disproved.

The body's mechanism for ridding itself of waste products consists of the kidneys, which are selective filters for the blood, and the pipe-and-storage system, which conveys the waste fluid, the urine, out of the body. This urinary apparatus, more than any other functional unit, suffered a long period of neglect and medical misunderstanding. For almost two thousand years it was Aristotle who had the best concept of its structure. Between his time and that of the Renaissance anatomists, the kidneys and their appendages were ignored, except for the completely fallacious attention paid to the appearance of the urine.

The genito-urinary system, in animals and in humans, was among Aristotle's multifarious interests. The first anatomical drawing ever known was prepared and described by him in *Historia Animalium*. The actual illustration is lost, but, according to Aristotle, it depicted the male testes, the urinary bladder, and parts of the ureters leading down from the kidneys.

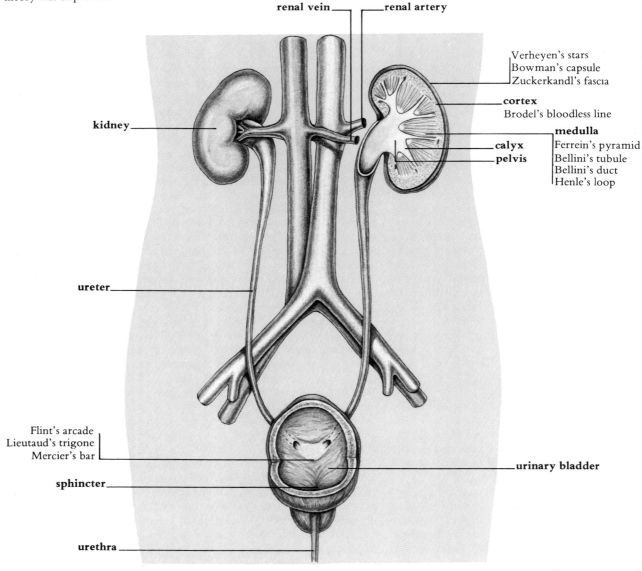

renal vein

renal artery

Verheyen's stars
Bowman's capsule
Zuckerkandl's fascia

cortex
Brodel's bloodless line

kidney

medulla

calyx
pelvis

Ferrein's pyramid
Bellini's tubule
Bellini's duct
Henle's loop

ureter

Flint's arcade
Lieutaud's trigone
Mercier's bar

urinary bladder

sphincter

urethra

Aristotle described the entire urinary system with clarity and in great detail. He explained the functioning of the *nephros*, the kidneys, and the "ducts from the aorta and the great vein," which were the renal arteries and vein, coming off and returning to the aorta and vena cava, which means hollow vein because it is usually empty after death. He called the bladder's trigone, the area where the ureters enter and the urethra exits, the "coalescence of the returning ducts" because the tubes from the male organs also terminate there. In essence, Aristotle's description was correct in every aspect and virtually no new knowledge was added for more than two millennia; on the contrary, much was lost.

Throughout the earlier part of the Middle Ages, one of the major preoccupations of physicians was the spurious science of urinoscopy. Numerous illustrations show earnest physicians, as well as unscrupulous quacks, holding up flasks of urine to study the colour and turbidity of their patients' water samples.

The procedure, which was but a refined form of magic, owed much of its origin to Isaac Judeus, an Egyptian doctor who lived from 845 until 940. A member of the Salerno school of medicine, he was physician to the rulers of Tunisia. Judeus was also a prolific medical author. He wrote on fevers, diet, drugs, and the examination of urine. His collected works were printed in 1515 and remained popular for more than one hundred years.

Judeus maintained that virtually every disease could be diagnosed by careful scrutiny of the urine. He recommended that the urine be studied meticulously for its colour, density, and the shape of the various clouds that often form in it. It is known today that these are merely precipitates of phosphates and other innocuous substances, but to Isaac Judeus they had the greatest import. He recognized four levels in the flask of urine, each corresponding to a part of the body. If the cloudiness was at the top, for example, the head was the seat of the disease, and so on down to the bottom layer which indicated trouble in the legs. This fanciful and ridiculous precept survived, and indeed flourished, for centuries.

To give Judeus some credit, he wrote a few far more practical hints for doctors, which have come down through Robert Burton's seventeenth-century book *Anatomy of Melancholy*. One of these hints was "Ask for your fee when the illness is at its height, because the patient will forget what you did for him as soon as he is better." Another was "Treating the sick is like boring holes in pearls, the physician having to be careful lest he destroy the pearl."

The cult of urinoscopy continued as long as it did because physicians had nothing more useful to put in its place. For centuries, patients would go to their doctors with their samples. Some-

times they had special wicker baskets in which to carry the flasks safely. Although today the examination of urine is commonplace and many a patient can be seen in the doctor's waiting room self-consciously clutching a gin bottle wrapped in brown paper, at least there is a rationale behind the testing, usually related to its chemical or hormonal content as revealed by sophisticated laboratory analysis. Medieval urinoscopy, however, was empirical mumbo-jumbo without a shred of logical justification. It was alleged that by examining the urine, a disorder of the "humours" could be observed.

The only possible excuse for perpetuating the practice of urinoscopy was the psychological effect on the patient, who quite probably was impressed by the antics of the physician as he assiduously examined the yellow fluid. This was borne out by Arnold of Villanova, the Spanish

Believed to be one of the earliest portrayals of urinoscopy, this bedside scene is one of a sequence of miniatures relating to the treatment of a sick woman which appeared in a late thirteenth-century medical manuscript. Here the physician has dropped the urine flask in alarm, indicating that from his observation of its contents the patient's life is at risk.

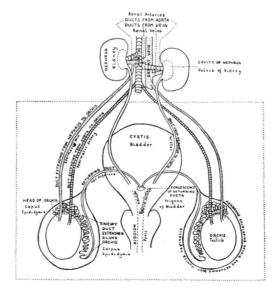

Aristotle did the first known anatomical drawing. It was of the male genito-urinary system. The area framed within the dotted rectangle is a restoration of the lost diagram. The terms Aristotle used are indicated in capitals, with the modern equivalents in italics.

Discovering the Human Body

In the early Middle Ages urinoscopy was regarded as an important aid to diagnosis. This woodcut, from Megenburg's *Book of Nature*, which was published in Augsburg in 1478, shows one physician holding the specimen flask while the other consults his notes.

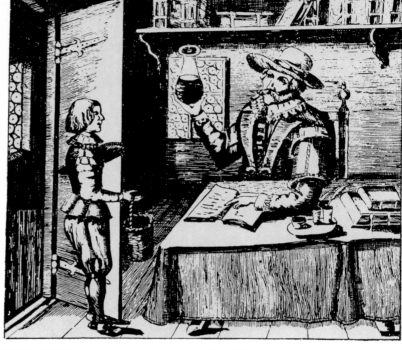

Inspection of urine also occurs in this scene at the bedside of a feverish patient, a woodcut by Hans Weinitz which appeared in the 1532 edition of Petrarch's *Mirror of Comfort*.

A seventeenth-century view of urinoscopy is shown in this engraving from Fludd's *Integrum Morborum Mysterium*. The boy at the door holds the special basket in which the urine flask was carried.

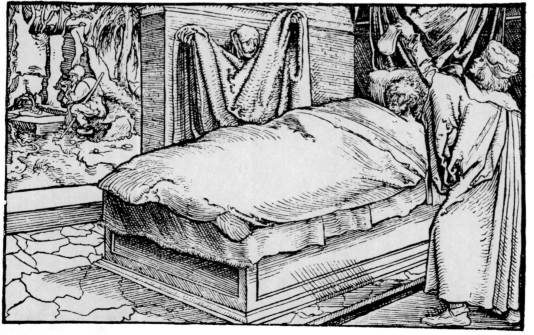

doctor who wrote in the thirteenth century: "If you find nothing wrong with the patient's urine, but he still insists he has a headache, tell him it is an obstruction of the liver. Continue to speak of obstruction, it is a word he won't understand, but it sounds important."

The school of Salerno obviously took this advice to heart. The Breslau manuscript of the Salernitans devotes no less than forty pages to the description of urine and alleged that no less than eighteen different colours of it could be differentiated.

After this ludicrous phase in medical history, there was little concern with the urinary system until the revival of descriptive anatomy. Vesalius's sixteenth-century book included excellent drawings of the kidneys and bladder, with the appropriate blood vessels and attached plumbing. The structure was more accurately dealt with than many other systems, undoubtedly because of its relative simplicity. No one ventured a guess, however, as to how the urinary system might function.

From the sixteenth century onwards, anatomists had a fairly good idea of the gross structure of the urinary system, but unlike the heart, the intestines, and the lungs, almost nothing further could be learned, or even guessed at, by mere naked-eye examination. The kidney was to be the organ of great interest to microscopists.

The basic unit of the kidney is the nephron, which consists of a small ball of capillary blood vessels, the glomerulus, pushed into the end of a long blind tube. This tube is looped and convoluted within the kidney until the lower end opens into wider ducts and then into the open space at the exit of the ureter, the pipe that takes the urine down to the bladder. The glomerulus can be thought of as a fist thrust into the outside of a plastic bag, turning it half inside out. The fluid part of the blood, the plasma, then leaks from the blood vessels through the bag and dribbles down the long tube which is fixed to the open neck of the bag. If that was all that happened, then the body's fluid would be rapidly lost. But, except in certain illnesses, the various substances needed by the body are reabsorbed by the walls of the tube, along with much of the water, so that only the unwanted wastes and excess water reach the other end and leak off into the bladder, from which, at convenient intervals, it is voided through the urethra.

The first great discovery was of the existence of the ball-shaped glomerulus, of which there are about one million in each kidney. Their discoverer was Marcello Malpighi, the Father of Microscopic Anatomy. He published his description of the glomerulus in 1659 in his book *De Viscerum Structura*.

The larger ducts, which collect the urine and void it into the pelvis, or open space, of the

kidney were described soon afterwards by Lorenzo Bellini and were henceforth known as Bellini's tubules. The ends of the tubules were called Bellini's ducts. This discovery was recorded in Bellini's book, which was published in Florence in 1662, when he was only eighteen.

When the surface of the kidney is examined, tiny, star-shaped veins can be seen scattered over the surface. These are Verheyen's stars, named after Philippe Verheyen who described them in 1699. Born in a rural hamlet in Belgium in 1648, Verheyen started to work on the land and then decided to become a priest. But one of life's curious accidents befell him. An injury resulted in one foot having to be amputated. This banned him from ordination and seemed to stimulate his interest in anatomy. He went on eventually to become a professor first of anatomy and then of surgery at Louvain.

Inside the kidney, the open space, the pelvis, where the urine collects, has side branches called calyces, or calyxes. The name is taken from their resemblance to the calyx of a flower. The substance of the kidney consists of an outer rim or cortex, where all the glomeruli lie, and the

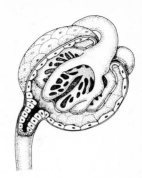

The nephron, shown here in an exploded view, is the microscopic filter unit of the urinary system. There are more than one million nephrons in each kidney. A highly specialized, coiled tubule, the nephron filters the kidney's blood supply to extract a concentrate of waste products which is carried down the ureter and evacuated as urine.

Marcello Malpighi was one of the earliest microscopists and he described things which had never been seen before. He was born near Bologna in 1628, the same year Harvey's momentous *De Motu Cordis* was published. Malpighi started to study philosophy at the age of seventeen and at about this time he began amusing himself with the microscope, which was then coming into use. In 1649, he began the study of medicine at Bologna. There he learned of the work of Harvey and was encouraged to become involved in medical research. In 1658, Malpighi was appointed to the chair of theoretical medicine at Pisa where he became a close friend of Gian Borelli. Four years later, because of ill health, he returned to Bologna, but left again for Messina in Sicily where Borelli had obtained a chair for him. Once more after four years Malpighi returned to Bologna, where he remained for the next twenty-four years. From the time that he began his own research, Malpighi's colleagues raised considerable opposition to his ideas and principles, which were contrary to the time-hallowed precepts of Galen. Malpighi's election to the Royal Society of England in 1669 was an honour that guarded his reputation against local prejudice, but his property and even his person were not as well protected; in 1684, his house was burned down and his microscopes and manuscripts destroyed. Five years later he was physically assaulted by two masked men who, it was discovered, were fellow doctors at the university. Even Borelli, who had been his mentor, turned against him and wrote vindictive disclaimers of his work. In 1691, Malpighi gave up his academic position and went to Rome at the invitation of Pope Innocent XII. Three years later Malpighi died.

Carl Ludwig, the physiologist, inventor, and teacher, who lived from 1816 to 1895, spent most of his life as a professor at Leipzig and his students called him "the greatest teacher who ever lived." In the physiological institute which he founded he devised equipment to improve research and was the first to keep animal organs alive outside the body. His descriptions of the function of the heart formed a basis for modern cardiology, and he explained the role of oxygen in the blood. He published much of his own work under the names of, or in collaboration with, his pupils.

William Bowman was born in Cheshire in 1816. He began his medical career as an apprentice surgeon at the Birmingham General Hospital and then continued his studies at King's College, London, where he later held the chair of anatomy and physiology. He became England's leading ophthalmic surgeon and was noted for his studies of the anatomy and physiology of the eye as well as of the kidney and urinary system. All his publications, including his great work *The Physiological Anatomy and Physiology of Man*, he illustrated himself. He died in 1892.

Friedrich Gustav Jacob Henle gained pre-eminence as a microscopic anatomist. His observations have been compared with the work of Vesalius. Born near Nuremberg in 1809, Henle studied medicine with Johannes Müller. He was imprisoned briefly for alleged political activities, and after his release he went to Zürich where he became professor of anatomy. Four years later he returned to Germany as professor of anatomy first at Heidelberg and then at Göttingen where he remained until his death in 1885. Henle, who anticipated the discovery of the cause of infectious disease, was a prolific writer and a gifted artist.

streaky medulla, where the tubules converge on the calyxes. Between the calyxes are the pyramids, sometimes called Ferrein's pyramids, after Antoine Ferrein, French professor of surgery and anatomy who described them in 1746.

Flint's arcade is a curious title for the arches of blood vessels that lie at the bases of the pyramids in the marginal zone between cortex and medulla. The name commemorates Austin Flint, an American professor of physiology who was born in Massachusetts in 1836 and was one of the founders of New York's Bellevue Hospital Medical College.

Today, the glomerulus is often likened to a cluster of blood vessels pushed into a plastic bag so that one layer of plastic is tightly applied to the vessels, with a narrow space between that layer and the other side of the bag. This double layer is known as Bowman's capsule.

William Bowman, one of the great surgeons of Victorian England, described this terminal capsule in a paper which was published in 1842 in the prestigious journal *Philosophical Transactions*. The article had a sonorous, but descriptive title, typical of that day: "On the Structure and Use of the Malpighian Bodies of the Kidney with Observations on the Circulation through that Gland." It was a classic description and one that had fundamental importance for physiologists and pathologists, because Bowman's capsule is a vital element in the filtration through the glomerulus and in some kidney disorders.

Not only did Bowman describe the capsule, but he was the first to recognize the relationship of the glomerulus to the tubules. He put forward the theory that water and perhaps salts pass out into the tubule. He had the erroneous idea, however, that water passed out in order to dissolve urea, the main waste product, which was ejected by the walls of the tubules, and so wash it out into the bladder.

The intermediate part of the nephron consists of a long, straight tubule with one bend in it, forming an elongated U, together with various tortuous parts understandably named the convoluted tubules. The long loop had to wait for a considerable time to be eponymized. It is known as Henle's loop because it was described by the German anatomist Friedrich Henle in his *Handbook of Systematic Anatomy*. Published over a number of years, the part relating to the kidney tubules did not appear until 1862.

Brodel's bloodless line lies between the areas of the kidney which get their blood from different branches of the renal artery and is of importance to those who operate on the kidney. It is named for Max Brodel who was a medical illustrator at the medical school in Leipzig, until he emigrated to the United States in 1893. He became director of the Medical Art Institute in Baltimore and is acknowledged as the founder of medical illustration in the United States.

Zuckerkandl's fascia lies behind the kidney and tethers it to the back of the abdominal wall. It commemorates Emil Zuckerkandl, an Austrian who was professor of anatomy at Graz and described the fascia in 1883.

In the urinary bladder, the trigone is the triangular area marked out on the floor of the organ by the orifices of the two inlet tubes, the ureters, and the outlet pipe, the urethra. It is sometimes called Lieutaud's trigone after Joseph Lieutaud, a French anatomist and pathologist who described it in 1742.

Another Frenchman is also represented in the bladder trigone. The ridge which joins the two ureteric openings, forming the rear boundary of the triangle, is known as Mercier's bar, described in 1848 by Louis August Mercier, a urological surgeon.

It was late in the eighteenth century that discoveries were made about the chemical nature of the constituents of urine. Urea, the main waste product derived from nitrogenous substances, was first identified in 1773. Carl Wilhelm Scheele, the great Swedish chemist, who also did much original work on oxygen, found uric acid in 1776 and within three years its implication in the disease of gout had been recognized.

Scheele had identified uric acid from the urine and in the same year it was found in calculi, the hard stones that form in the kidney and bladder due to precipitation of the solid constituents.

The stone, as it is universally called, was one of the few urinary diseases known in antiquity; it is mentioned in the Hippocratic Oath. Proper physicians did not stoop to operate, but left this to the mere surgeons. Bladder stones were the major stock in trade of civil surgeons for many centuries and were part of the reason physicians considered surgeons inferior. All manner of unqualified people, including barbers, would "cut for the stone." It was speed and accuracy, rather than clinical acumen, which distinguished a stone-cutter in those days.

It was the brilliant nineteenth-century Viennese physiologist Carl Ludwig who worked out much of the basic function of the kidney. He had done work on blood pressure and had formulated the hypothesis that the glomerulus is a passive filter, dependent on the blood pressure to drive the water and salts through Bowman's capsule. This fluid, Ludwig suggested, is then concentrated by osmotic pressure in the tubules, so that what arrives at Bellini's ducts is much stronger in salt concentration than the original plasma. The cells and large protein molecules in the blood he believed are held back by the glomerulus.

Ludwig put these ideas forward in 1844. A

Circumference of the Stone 8 Inches

decade later they were substantiated by other investigators who found that the rate of filtration, as the passage of fluid across the glomerulus is called, is proportional to the blood pressure. But Ludwig was not satisfied with his concept because he knew that the concentration of different substances in the urine is different from the original ratio in the blood, which ruined a simple filtration-absorption theory.

In 1869 he discovered that it is not only the blood pressure that affects filtration, but also the concentration of substances in the blood. Five years later, however, Rudolf Heidenhain, the German physiologist, and professor of histology and physiology at the University of Breslau, found experimental evidence which tended to return to Bowman's original concept. For a considerable period there was a controversy between the Ludwig and the Heidenhain theories.

The subsequent history of the understanding of renal function becomes extremely complex. In the later years of the nineteenth century, delicate techniques were devised to investigate the function of various parts of the nephron. More general tests made it possible to discern how various chemicals are eliminated from the bloodstream. These procedures helped to gain the comprehensive knowledge that exists today, which is a far cry from diagnosis according to eighteen different colours of urine employed by the medieval urinoscopists.

Removal of a bladder stone was one of the earliest surgical procedures. This bladder stone, which was eight inches in diameter, was removed at Guy's Hospital in London in 1739 from a young man who "obtain'd a perfect cure."

The Brain and Central Nervous System

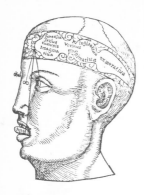

This portrayal of the cell doctrine, which postulated the cerebral ventricles as the central core of intelligence, was first published in 1503 in Gregor Reisch's *Margarita Philosophica*.

The nervous system is the most complex mechanism in the body and the story of how it came to be understood is equally complicated. The system is conventionally described in two major parts—the brain itself and the rest of the nervous tissues which consist of the spinal cord, the autonomic nervous system, and the peripheral nerves. This is really an artificial separation because the nervous system is an excellent example of total physiological integration, a mass of interrelated reflexes and pathways which make the most sophisticated electronic computor look like an abacus by comparison.

When tracing the history of the discovery of the nervous system's structure and function, the brain cannot be realistically separated from the nerves. And without an overview of the whole system, the achievements of any one pioneer cannot be adequately measured. With such a mass of intermeshed data, the only hope of creating some order is to work on a chronological scale, going back to early days of history.

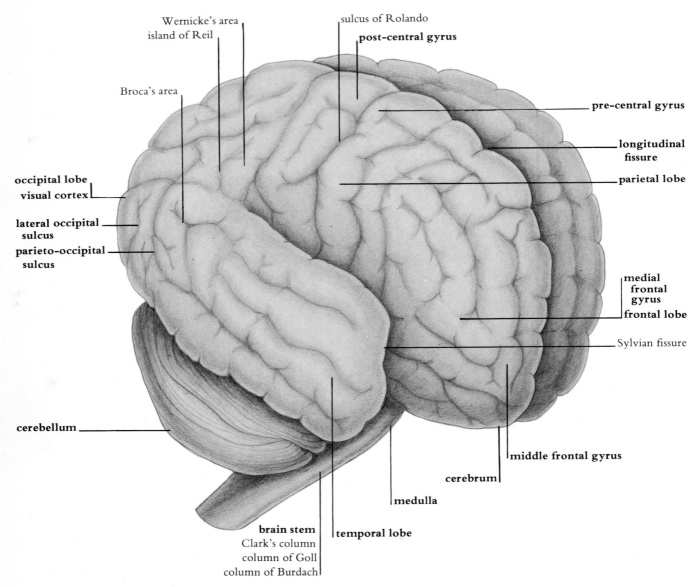

Wernicke's area
island of Reil
sulcus of Rolando
post-central gyrus
Broca's area
pre-central gyrus
longitudinal fissure
occipital lobe
visual cortex
parietal lobe
lateral occipital sulcus
parieto-occipital sulcus
medial frontal gyrus
frontal lobe
Sylvian fissure
cerebellum
middle frontal gyrus
cerebrum
medulla
brain stem
Clark's column
column of Goll
column of Burdach
temporal lobe

The ancient Egyptians seem to have paid scant attention to the nervous system, being content to extract the brain piecemeal through the nostrils when preparing their dead for embalming. Herodotus, the Greek Father of History, wrote that the Egyptians did not regard the brain as being of any importance. They thought that the seat of the soul was in the heart and the liver and this belief was accepted by some famous Greeks. The Edwin Smith papyrus, a medical scroll discovered in 1862, mentions the brain, however, and even includes such details as the presence of cerebral convolutions and meninges, the coverings of the brain. Paralysis of the bladder and intestines is mentioned in relation to spinal cord damage and it is said to indicate that the brain was the site of mental functions.

Whatever the truth about Egyptian beliefs might be, it is with the Greeks that well-documented awareness of the central nervous system begins. Even here, basic concepts swung wildly back and forth for a few centuries.

Alcmaeon of Croton was one of the first to mention the brain and in the late sixth century B.C., he stated that it is in the brain, and not in the heart, that the seat of the senses and the centre of intellectual life should be sought. Alcmaeon believed that sleep was due to a temporary retreat of blood from the brain—and that death was due to the blood's permanent exodus.

In about 450 B.C., Plato stated that the brain was the home of the intelligence, using as supporting evidence the fact that the spherical shape was an ideal residence for reason. Hippocrates, the father figure of medicine, agreed with this view. In a treatise on epilepsy, which was known as the "sacred disease," it was mentioned that "it is from the brain and from the brain only that our feelings arise." Hippocrates, or whoever of the Hippocratic school actually wrote the book, recognized the two symmetri-

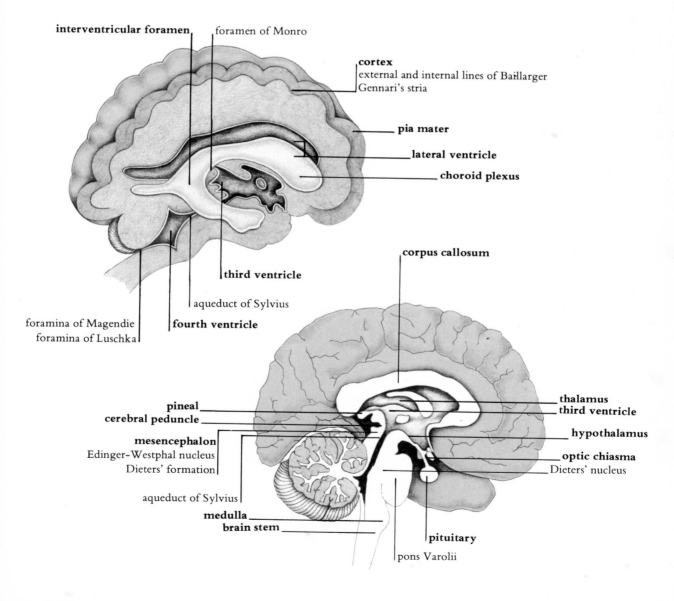

interventricular foramen
foramen of Monro
cortex
external and internal lines of Baillarger
Gennari's stria
pia mater
lateral ventricle
choroid plexus
third ventricle
aqueduct of Sylvius
fourth ventricle
foramina of Magendie
foramina of Luschka

corpus callosum
thalamus
third ventricle
hypothalamus
optic chiasma
Dieters' nucleus
pineal
cerebral peduncle
mesencephalon
Edinger-Westphal nucleus
Dieters' formation
aqueduct of Sylvius
medulla
brain stem
pituitary
pons Varolii

cal halves of the brain and the blood supply, but extraordinarily claimed that one blood vessel came from the spleen and the other from the liver. Time and translators may have confused the meaning. Probably what was meant is that there is one vessel on the right and one on the left, corresponding to the sides where the spleen and the liver are situated.

The pre-eminence of the brain suffered a sharp reversal with the pronouncements of Aristotle a century later. He discounted the importance of the brain and replaced the seat of the soul in the heart, where the Egyptians had sited it. Aristotle believed in a *sensorium commune*, a phrase which has persisted as "common sense," although in ancient times it meant a generalized function of intelligence, without localization.

Although Plato in *Timaeus* and Aristophanes in his play *The Clouds* had established the brain as the receptacle of consciousness, Aristotle was apparently aware that the brain is devoid of sensation and so concluded that it could not be the home of feelings of any kind. He felt, therefore, that it was merely a device for cooling the heart, the organ he considered most important.

The more practical Greeks of Alexandria, who used dissection of humans as well as animals in their search for information, came up with the first hard facts about the structure of the nervous system.

Herophilus of Chalcedon, the premier anatomist of the Ptolemaic medical school at Alexandria, firmly reinstated the brain as the receptacle of the intelligence, regarding it as the centre of the nervous system. He was the first to appreciate the true nature of the nerves as the communications network of the body. Furthermore, he grasped the dual nature of the nervous trunks, some being responsible for the propagation of muscle commands, while others are the vital transmitters of sensation. Although Hero-

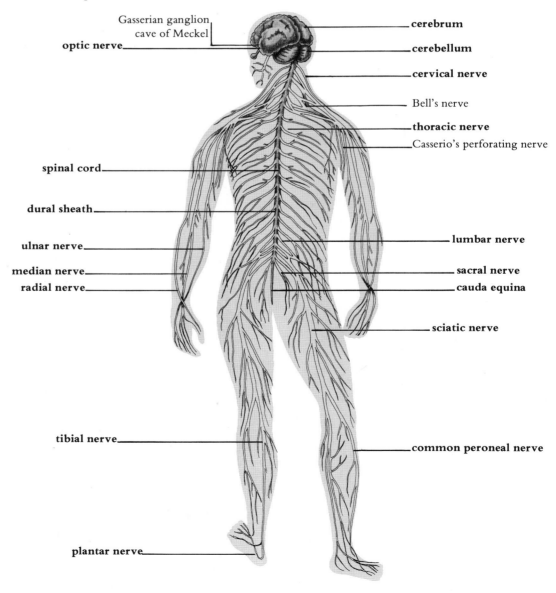

Gasserian ganglion
cave of Meckel

optic nerve

cerebrum

cerebellum

cervical nerve

Bell's nerve

thoracic nerve

Casserio's perforating nerve

spinal cord

dural sheath

ulnar nerve

lumbar nerve

median nerve

sacral nerve

radial nerve

cauda equina

sciatic nerve

tibial nerve

common peroneal nerve

plantar nerve

philus used the word neuron, he had no concept of its meaning in the modern sense and included ligaments and tendons in this category. As the result of his dissections, he recorded the meninges, the coverings of the brain, and distinguished the major parts of the brain itself, dividing the organ into cerebrum and cerebellum. It was Herophilus who first mentioned the ventricular system, which was to assume such great—and incorrect—importance in medieval times.

Another misconception, as far as human anatomy and physiology is concerned, was Herophilus's description of the rete mirabile, the meshwork of vessels at the base of the brain which exists in some animals, but not in humans. Galen was to perpetuate this error and mislead progress in that direction as in so many others. Still, a great debt is owed to Herophilus for his first descriptions of the brain and spinal cord and for distinguishing nerves from blood vessels, a differentiation that was blurred in the Hippocratic writings.

Erasistratus of Chios also advanced neuroanatomy and physiology at Alexandria. Like Herophilus, he was quite definite about the central role of the brain in consciousness. He was a rationalist and believed that nature moulded animal structures to suit their environment, rather than accepting Aristotle's view that structure was an innate property due to the soul, or psyche.

Erasistratus confirmed that the brain is divided into various prominent parts, including the cerebral hemispheres and the cerebellum. He gave a more detailed description of the internal ventricular system, the series of chambers that are connected deep in the substance of the brain where the cerebro-spinal fluid is manufactured by the vascular choroid plexus. He noted that the sulci and gyri, the ridges and furrows, on the surface of the brain are more prominent in humans than in lower animals, and he correctly associated this with the differing levels of intelligence.

Erasistratus traced the nerves towards the brain and spinal cord and at first wrongly assumed that they ended in the dura mater, the thickest of the three membranes which clothe the brain and cord. Later he made more detailed dissections and realized that the nerves penetrated the dura and continued into the actual nervous substance beneath. Like Herophilus, Erasistratus distinguished between motor and sensory nerves, those which carry messages for movement and those which bring sensation from the periphery.

Perhaps Erasistratus's most noteworthy contribution was the most fallacious, for it was the theory which led Galen into his fantasies that were then imposed on the world for a millennium and a half. This was the concept that the

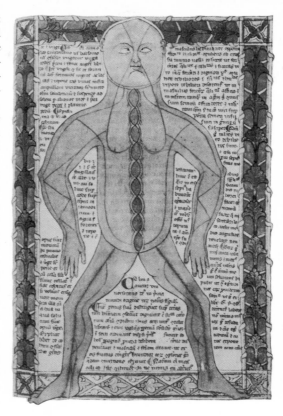

In this crude and simplistic illustration of the nervous system no attempt was made to integrate the system as a whole. It is from a thirteenth-century English medical manuscript.

ventricles of the brain were filled with some mystical substance called animal spirit which, several centuries later, Galen seized upon for his own hypothesis.

Yet Galen cannot by any means be dismissed as a mere purveyor of unlikely theories. He carried out a number of experiments which were pertinent to an understanding of the nervous system. He showed, for example, that if the spinal cord was slit longitudinally little abnormality resulted, yet if it was cut horizontally total loss of sensation and movement occurred below the section. When half the cord was cut across, then loss of muscle movement took place only on that side. He realized that the nerves coming from the cord have a segmental distribution; they supply the area in horizontal bands around the body related to the level of their origin in the spinal cord.

The phrenic nerves, which control the diaphragm (and cause hiccoughs), were described and understood by Galen. He also dissected out the laryngeal nerves which supply the vocal cords. He showed that by tightening the noose of a ligature on the nerves, the cries of an animal could be stopped, the sounds returning when the thread was loosened. Consequently, a branch of the superior laryngeal is known as Galen's nerve. Galen's experiment, however, led to a long-standing misapprehension that the nerves worked by hydraulic pressure. When the ligature was tightened, Galen assumed that the

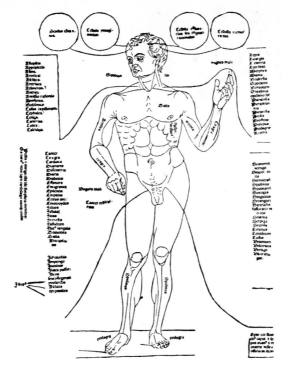

A Renaissance drawing of a Disease Man, from the illustrated medical book *Fasciculus Medicinae* by Johannes Ketham, which was published in 1491, perpetuates the medieval cell doctrine. The elements of intelligence appear in four balloons which represent the ventricles of the brain.

internal hydraulic fluid was interrupted, especially because the function returned when the pressure was released.

Galen also insisted that there were only seven cranial nerves—nerves which arise directly from the base of the brain rather than from the spinal cord. There are actually twelve, but his magic figure of seven was accepted throughout the Middle Ages.

It was Galen's general hypothesis of human physiology that was particularly tortuous and obscure. It involved the cardio-vascular and respiratory systems and its integrated nature drew the nervous system into the fold as well. According to Galen, the nervous impulse was operated by some hydraulic-type substance passing down the "hollow" tubes which were the nerve fibres. This substance he declared to be animal spirit which was formed in the ventricles of the brain by vital spirit, carried by the blood and changed in some occult way in which the nonexistent rete mirabile network at the base of the brain was somehow involved. The animal spirit then coursed down the hollow nerves and operated the muscles and conveyed sensation.

This extraordinary theory of nervous function was accepted for century upon century by generations of doctors equally as intelligent as those today. Yet the psychological constraints of the time were such that even in the seventeenth century learned physicians did not even consider challenging the sacred thoughts of Galen. Thomas Willis of London, for example, the man after whom the brain's circle of Willis

is named, declared that the observed loss of brain function which takes place when the cerebral blood supply is interrupted, resulted because "vital spirits could not reach the brain for conversion into animal spirits." As late as 1671 he repeated Galen's experiment of ligating nerves and showed how the ensuing paralysis was due to obstruction of the flow of animal spirits.

This animal spirit was sometimes called the psychic spirit by Galen and he considered that it was concerned with intellectual activities as well as with sensation and movement. These intellectual activities included imagination, memory, and thought, and, in contradiction of Aristotle, Galen did believe that the different brain functions were located in specific parts of the brain.

Although Galen's theories about the brain left much to be desired, his knowledge of the spinal cord was of a higher level. He sectioned the cord to observe how the loss of function varied and this was elaborated to give much more information about the function at different levels. He discovered that cutting across the cord in the neck region gave results which varied from sudden death to increasing respiratory difficulties. When the lower part was severed, paralysis of the lower limbs, bladder,

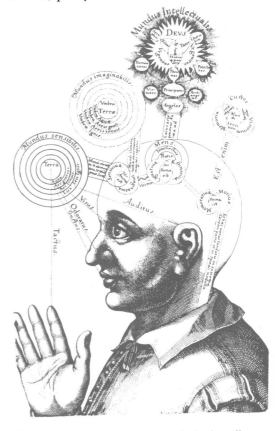

This seventeenth-century diagram links the cell doctrine with spirituality and intellectualism. Where cells representing the cerebral ventricles overlap are written the words "The soul is here."

and intestines occurred. Little more was learned until the nineteenth century; therefore all Galen's work can by no means be dismissed as merely speculative nonsense.

For centuries, the ventricular system inside the brain was regarded as the vital part of the apparatus. The white and grey matter which makes up by far the larger mass was disregarded as having a subservient role. Herophilus described four ventricles and stated that the soul resided in the fourth, which was placed at the back of the others. To give Galen his due, he was one of the few who thought that the solid part of the brain was the more important, but after his time the ventricles, now said to be three in number, became firmly established as the central core of the intelligence.

The reduction in number from four to three was because medieval anatomists combined the right and left lateral ventricles, leaving the narrow third and the posterior fourth as the remaining cells. Scores, if not hundreds, of drawings and descriptions appeared in the following centuries, depicting a head with either three spherical cavities or three rectangular divisions. These usually had speculative labels attached, giving the function of each cell in terms of intellectual activity, special sensation, and motor function. This cell theory persisted for hundreds of years.

Mondino included a similar plan in his book published in 1316. The double front cell was, he said, the meeting place of the senses. The middle cavity was the seat of imagination, and the posterior cavity the seat of memory. According to Mondino, mental function was controlled by a "red worm," by which he meant the fleshy choroid plexus that lies in the ventricles. He considered that it controlled brain function by moving hither and thither, blocking off the narrow communications between the ventricles, so modifying the flow of the animal spirits. Mondino still believed, according to the old Aristotelian principle, that the brain cooled the heart, although he hedged his bets by accepting Galen's view that it was also the seat of consciousness and intelligence.

One of Leonardo's drawings also illustrates the three-cell doctrine, the front cell linked to the eyes. But Leonardo da Vinci was one of the first Renaissance artists to draw the brain accurately; he made wax casts of the ventricular system, drawings of which are in his notebooks.

In the sixteenth century, researchers began to be bold enough to deny Galen. Berengar of Carpi, who was professor of surgery at Bologna between 1502 and 1527, wrote a commentary on Mondino's work which contained many original observations of his own. He denied the existence of the rete mirabile in humans, as described by Mondino, who himself was commenting on Galen.

Jacob Berengar gave a better account of the anatomy of the brain than anyone to date, albeit in excruciating medieval Latin. He described the ventricles, the choroid plexus, which is made up of arteries and veins, the pineal gland, considered by some to be the seat of the soul, and the fourth ventricle.

Andreas Vesalius's work on the nervous system, as displayed in his famous *De Fabrica* of 1543, was not as good as much of his other anatomizing, but it was systematic and continued the trend of such men as Berengar in disposing of the cell theory. To Vesalius, the ventricles were merely anatomical parts, with no intrinsic function. Still a disciple of Galen, he undermined the old philosophy unwittingly, explaining that he was brought up on the cell doctrine, but as the result of his descriptive talents he at the same time reduced its mysteries.

In *De Fabrica*, the numerous illustrations of the brain as seen from above are excellent. The brain,

Although it is nonexistent in humans, the rete mirabile was involved in Galen's humoral concept and was believed to convert vital spirit into the more subtle animal spirit, which then flowed through the nerves as a force for movement and sensation. Even Vesalius accepted the existence of the rete mirabile in humans and included it, marked B, in this illustration of the arteries which appeared in his *Tabulae Anatomicae Sex* published in 1538. The only anatomist up to this time to deny its presence in humans was Jacob Berengar of Carpi.

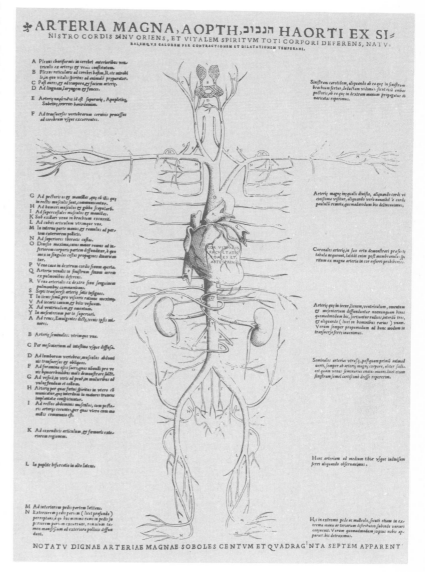

In *De Fabrica,* Vesalius's drawing of the mythical rete mirabile, above, had the pituitary gland at its centre. In the figure right, the moustache drawn in as a touch of Vesalian whimsy, overlying tissue has been sectioned back to reveal the great vein of Galen (H) running between the lateral ventricles. Vesalius included a view, far right, of the base of the brain, "so that the origin of the cerebral nerves may be suitably exposed to view." In it the cranial nerves are classified—incorrectly—according to Galen.

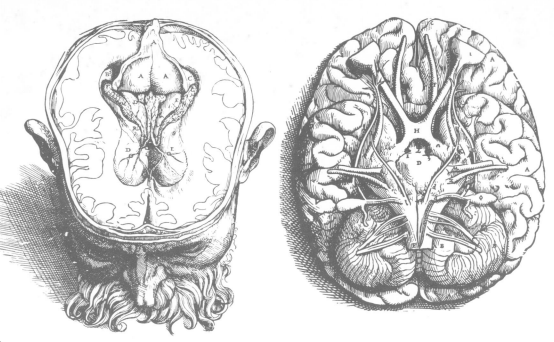

still within the skull, is cut across in different horizontal planes. And yet, Vesalius still included the rete mirabile, which could not have been the result of observations of human dissection. Even Leonardo included it in a drawing of the base of an indisputably human brain. Because Galen had said it was there, it must be there, even if it was invisible. Galen's dictate overrode the common sense of Vesalius and Leonardo da Vinci. But at least Andreas Vesalius changed his mind. Between his *Tabulae Sex* of 1538 and *De Fabrica* of 1543, he dispensed with the rete mirabile in his text, yet strangely enough persisted in including a drawing of it, even in the later book.

Vesalius continued to perpetuate Galen's seven cranial nerves. His rendering of the peripheral nerves is not outstanding and it is his rival Eustachius who is renowned for a magnificent plate of the sympathetic nervous system. This was not published until long after his death, but it remains one of the best representations of the system even to this day.

As to knowledge of the function of the brain and nervous system in those times, Vesalius repeated the Galenic experiments of ligating nerves and cross-sectioning the spinal cord. As did Leonardo, he recognized the reciprocal action of muscles—that when a muscle contracts to perform a movement, the opposing muscle must relax. Vesalius removed the sheath from a nerve and showed that it was stimulation of the exposed end that made a muscle act.

From this point on, the modern age of neuroanatomy and physiology can be said to have arrived. Galen's theories were progressively discarded, although many of his tenets were clung

to for an amazingly long time, certainly well into the seventeenth century.

The brain and spinal cord are sheathed by the meninges—a triple layer of membranes which all the nerves must penetrate to reach the rest of the body. The outermost layer is the dura mater, a tough fibrous sheet which lines the inside of the skull and spinal canal. Its name, which means strong mother, is evocatively apt for this protective layer.

Inside the dura mater is the arachnoid mater, a delicate meshwork which takes its name from the spider's web it resembles. The arachnoid carries most of the blood supply and beneath it is the layer of cerebro-spinal fluid which bathes the brain and acts as a hydraulic buffer against shocks and stresses.

The innermost membrane is actually the outer covering of the brain and cord and is inseparable from it. So closely does it cling that it is called the pia mater, the pious mother.

Where each blood vessel dips from the arachnoid into the substance of the brain a sheath of inner membrane follows it, leaving a potential space between the membrane and the vessel. This is the Robin-Virchow space and is considered by some anatomists to be identical to a lymphatic channel, although no true lymphatics exist within the central nervous system. The two names are independent of each other, but have been combined to avoid any

In 1552, Eustachius assembled a series of copperplate engravings, including this excellent drawing of the base of the brain and the sympathetic nervous system. They were not published, however, until 1714.

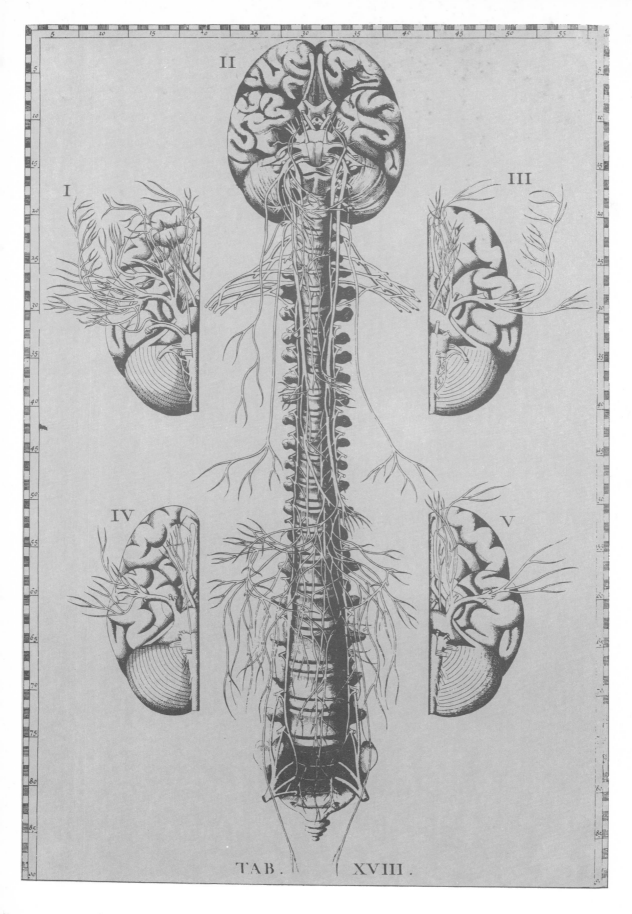

TAB. XVIII.

dispute about who was the first to describe the space. Charles Filippe Robin was a professor of histology in Paris and Rudolf Virchow was the great nineteenth-century pathologist. Both men wrote about this space in the 1850s.

The brain itself is composed of three main parts—the cerebrum, the cerebellum, and the brain stem. It is supported within the skull by tough extensions of the dura mater.

The cerebrum is divided into two great halves—the cerebral hemispheres—and these are separated in the mid-line by a large crescent of membrane, the falx cerebri, so named because of its sickle shape, which sweeps down from the roof of the skull.

The front two-thirds of the base of the cerebral hemispheres is supported by the floor of the skull. But at the back, where the softer cerebellum lies beneath, a taut, drum-like membrane crosses from side to side. It is held up in the middle by the bottom edge of the falx and looks rather like a shallow ridge tent. Not surprisingly, it is called the tentorium.

Joining the two halves of the cerebrum is the large bridge, the corpus callosum, which carries fibres from one side to the other. Galen described it as a firm, tough body and corpus callosum is merely a translation of Galen's term. It first appeared in print in the sixteenth century in Vesalius's *De Fabrica*.

The pineal, which means pine cone, is a rather mysterious nodule under the back end of the corpus callosum. It is now thought to be an endocrine gland with some function relating to the generative organs. Some anatomists, however, believe that it is a remnant of a third central eye which humans had in some distant phase of evolution.

The cerebral cortex, the convoluted grey matter which covers the surface of the cerebral hemispheres, is the most sophisticated part of the brain. Human mental superiority is due to its large area and complex structure. The ridged and furrowed surface of the human brain permits a large surface area to be accommodated in the smallest space, just as a crumpled piece of paper can provide a large surface in a small volume.

Although this complex crinkling was first related to human intelligence by Erasistratus more than two thousand years ago, the surface area is only part of the story, for the complexity of the layers, or strata, within the cortex is fantastic. Layer after layer of cells of different types and sizes, together with their connecting fibres, pass downwards from the pia mater until the white matter, which serves purely as connecting cable, is reached.

The grey matter of the cortex, together with the large basal ganglia and other scattered areas of nerve cell tissue, is the part which actually initiates all the activities of the brain. In different areas of the cortex this structure varies according to the functional needs. In the occipital regions at the back of the brain, for example, the areas concerned with sight have layers with special names. One of these is Gennari's stria. A stria is a streak or layer, and Francesco Gennari, an anatomist of Parma, described the layer in 1782.

In 1840, a Parisian psychiatrist, Jules Gabriel François Baillarger, showed that Gennari's line was really double and he named two lines, the outer of which was said to be continuous with the line of Gennari and the other beneath it. The two white lines or bands described by Baillarger have since been known as the external and internal lines of Baillarger.

On each side of the cerebral hemisphere there is a prominent groove, or sulcus, which runs obliquely downwards and forwards at about the centre of the brain surface. This is the central sulcus, or fissure, of Rolando, which was named after Luigi Rolando, a professor of anatomy at the University of Turin early in the nineteenth century.

Rolando's fissure separates two large convolutions, or gyri; the front one, the pre-central gyrus, is the motor centre of the body which initiates almost all the movement of the voluntary muscles. Behind the fissure is the post-central gyrus, which receives most sensory stimuli, except from the ears, eyes, and nose, which have special areas elsewhere in the brain.

The two large gyri on each side are the centre of appreciation of touch and position and of

Pierre Paul Broca, the French surgeon and anthropologist, was born in 1824. Although he was interested in higher mathematics he graduated in medicine and became a professor of clinical surgery in Paris. He studied the pathology of cancer and the treatment of cerebral aneurism and discovered the speech centre in the third frontal convolution of the brain which is now known as Broca's area. Broca became interested in anthropology, devised many instruments and methods of measuring skulls, and was involved in the discovery and study of Neanderthal man. In 1859, Broca founded the Society of Anthropologists in Paris, which was regarded with great suspicion by the government. The society was permitted to meet only if Broca accepted personal responsibility for all that was said "against society, religion, or the government." After his death in 1880, the Musée Broca was established. On display there was his collection of craniums from all over the world which he used in his studies of comparative brain structures and functions. It now forms part of the Musée de l'Homme.

voluntary movements. To help visualize the location of the body image on the gyri of the cortex, students and doctors remember two diagrams, each with a misshapen little man, a homunculus, lying along it. Each part of the homunculus corresponds with the part of the body connected with the nerve cells of that point in the gyrus.

In the motor cortex, the homunculus has a huge face, lips, and tongue and a gigantic hand with an enormous thumb. This indicates the dominant parts supplied by the motor nerves, the mobility of the opposable thumb being the hallmark of humanity's success, since the ability to grip and perform delicate manipulations marked humans off early from other animals. The power of speech also demands a large cortical representation for the lips and tongue. The leg and foot, with little in the way of complex movements, are small appendages hanging over the top of the gyrus and down into the central fissure.

The sensory homunculus is of a different shape, although it, too, hangs upside down over the side of the brain with its foot over the top where, incidentally, the appreciable sensation from the genitalia is represented. The largest area is again for the lips, with tongue and thumb prominent, indicating the zones from which the most sensitive stimuli are obtained.

The motor zone for speech is lower down on the side of the frontal lobe and is known as Broca's area. Pierre Paul Broca, an eminent Parisian anatomist and surgeon, has several brain eponyms, all relating to his discovery of the speech area.

Broca's area is almost always on the left side of the brain when individuals are right-handed. About 14 per cent of people are left-handed and they have their speech motor area on the right. There is no satisfactory explanation for this dominance of the right hand and arm, nor for the prevalence of left-handedness in men; women are left-handed only half as often as men are. When a naturally left-handed person is forced to use his or her right hand instead of the left there is a risk of producing a speech impediment as the result of interference with Broca's area.

While the production of sound is controlled by Broca's area on the lower part of the frontal lobe, the reception and appreciation of sound from the ears is located on the upper gyrus of the temporal lobe in Wernicke's area. Karl Wernicke was a famous German neurologist in the latter half of the nineteenth century. A professor at Berlin, Halle, and Breslau, he is also eponymously remembered for a neurological syndrome known as Wernicke's encephalopathy and for publishing an excellent atlas of the brain.

At the side of each cerebral hemisphere, the cortex dips inwards and then comes out again,

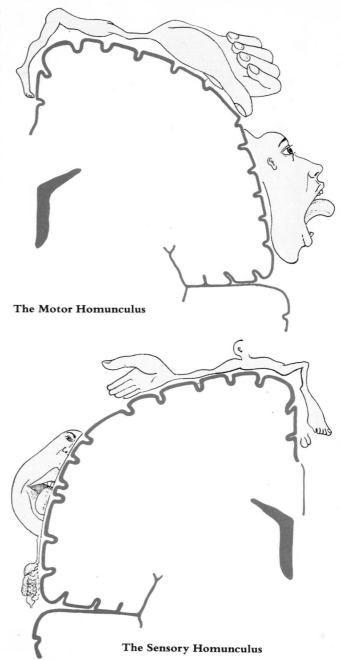

The Motor Homunculus

The Sensory Homunculus

In 1957, Wilder Penfield and T. Rasmussen, at the University of Montreal, devised the motor homunculus and the sensory homunculus superimposed around the profile of one hemisphere of the brain.

forming a T-shaped cleft. The area of cortex hidden in this bay is known as the insula, or island, of Reil. It is named after Johann Christian Reil, a German physician and professor of medicine who described it in 1796.

Inside each cerebral hemisphere is a large, fan-shaped spread of nerve fibres passing upwards and downwards from the brain stem to the cortex. This is the corona radiata, the spreading crown. The lower part, which carries the motor fibres to the muscles, passes through a gap between parts of the basal ganglia. At this point, the radiation of fibres is called the internal capsule and is buried deep within the brain.

The region of the external capsule is served by rather fragile blood vessels, one of which is Charcot's artery, the artery of cerebral haemorrhage. If this artery becomes blocked by thrombosis, then a cerebral haemorrhage, a stroke, will ensue. This transmits pressure on the internal capsule. The effect will depend on the part of the internal capsule which becomes damaged. Because the motor fibres cross over in the medulla oblongata in the brain stem, if the haemorrhage is towards the front of the right side of the internal capsule, for example, then weakness of an arm can be expected. When the damage is at the rear end of the left side of the capsule, paralysis of the right leg will occur.

When looked at from underneath, the brain has a deep central depression, the basal cistern, which is filled with the clear cerebro-spinal fluid, a protein-free filtrate of blood plasma produced by the choroid plexus inside the brain ventricles. The basal cistern is a protected area under the brain, where the stalk of the pituitary gland, the master endocrine gland of the body, passes down from the brain.

Around the cistern is the circle of Willis, the ring of vessels which provides an alternative pathway for blood to the brain should one of the four main arteries become blocked or damaged. In 1664, Thomas Willis described the circle in his book *Cerebri Anatomie*.

From the cistern the two deepest grooves in the brain run off on each side, partly dividing each cerebral hemisphere into two, separating the lower temporal lobe from the frontal and parietal lobes above. This deep division is known as the Sylvian fissure. It is named after François de la Bois, the seventeenth-century professor of medicine in Leyden who was known as Sylvius. He described the fissure in 1660, although it had already been noted in 1641 in a book edited by Caspar Bartholin.

Throughout the central part of the brain there are a number of masses of grey matter which are the nuclei of various nerve tracts. They may be thought of in some cases as relay stations or junction boxes, although actually they have a vital role in brain function on a lower level of consciousness than the cerebral cortex. Part of the function of the largest of the masses, the thalamus, for example, is the appreciation of pain.

The nuclei in the depths of the cerebrum are called basal ganglia, or corpora striata, meaning streaky bodies. Curious names attach to some of them. Thalamus is a Greek word meaning bridal chamber or bedroom. The lentiform nucleus nearby is named after a lentil bean because of its shape. The thin claustrum's name means a lock,

and the globus pallidus is a pale ball. A number of structures are named after fruit and vegetables. There are two olives as well as the amygdala, which means almond. The putamen is a fruit stone or shell.

Two little round nuclei which project from the surface of the basal cistern are known as mammillary bodies because of their resemblance to breasts. The calamus scriptorius, which means reed pen, is a small nucleus first described by Herophilus in the third century B.C.

Another well-known brain nucleus is the Edinger-Westphal nucleus, which is part of the control centre for the third cranial nerve which moves the eyes. Two men share the eponym for this important centre, although they worked independently. Ludwig Edinger was a German anatomist and professor of neurology at Frankfurt-am-Main. He described the nucleus in a medical journal in 1885, the same journal in which two years later Karl Westphal, a noted neurologist, director of the Berlin Brain Institute, and professor of psychiatry, announced his findings.

Another important nucleus is part of the eighth nerve complex. This nerve is really part of two separate systems, one related solely to hearing and the other to the vestibular system which controls balance. The two parts run together since the organs of hearing and balance are both inside the dense bone of the skull's petrous temporal.

The brain nucleus connected to the vestibular branch of the eighth nerve is Dieters' nucleus and is named after Otto Dieters, who was a pupil of the great Virchow. A professor of anatomy in Bonn in the middle of the nineteenth century, Dieters's name is associated with a number of nerve structures.

The base of the brain, near the mammillary bodies, carries the stalk of the pituitary gland, which leads down from the infundibulum, meaning a funnel, and is so named because of its shape. This is connected above to the important and evolutionarily ancient hypothalamus, its name a topographical description, for it means below the thalamus.

In front of the infundibulum lies the optic chiasma, the prominent X-shaped junction of the optic nerves and tract. Chiasma is Greek for crosswise arrangement, which accurately describes the crossing over of half the nerve fibres from the retina of the eye, so that all the images falling on one side of each eyeball go to the same side of the brain.

Within the brain and spinal cord there is a central cavity of complex shape which forms the ventricles and central canal. (Ventricle literally means little belly and many cavities in the body bear this name.) In the central nervous system the cavity forms in the embryo. The brain and spinal cord develop from a thickening

Constanzio Varolio, a professor of anatomy at Bologna, found a new way to dissect the brain, starting from the base rather than from the dome. His illustration of the base of the brain was published in 1573. The drawing is not as good as that of Eustachius, but Varolio is credited with being the first adequately to describe and draw the pons, which is still known as the pons Varolii.

surface ridge in the early foetus. This rolls over into a tube, leaving a channel down the middle. At the brain end, this cavity expands into a complicated system of chambers, which was revered by ancient and medieval anatomists as the true seat of mental activity. This gave rise to the cell doctrine persistent in the Middle Ages.

In each cerebral hemisphere there is a dolphin-shaped lateral ventricle. Each of these joins a third ventricle in the centre of the brain by means of a hole called the foramen of Monro. It commemorates Alexander Monro the Second. Three Alexander Monros occupied the chair of anatomy at Edinburgh in succession, covering a period of more than one hundred years. Alexander Secundus described the foramen in 1797, but said that he noted it forty-four years earlier. His son, Alexander Tertius, gave his father credit for the discovery in his own textbook of 1825.

The cerebro-spinal fluid, which is colourless and often described as "gin clear," is made in the ventricles by the choroid plexuses. It passes through the foramen of Monro into the slit-like third ventricle, which lies behind a membrane called the septum pellucidum, the transparent fence. It then flows through a narrow tube in the mid-brain, the aqueduct of Sylvius, which is named after the same Sylvius who is associated with the Sylvian fissure.

In the next chamber, the fourth ventricle, which is hidden under the cerebellum, the fluid can escape into the sub-arachnoid space where it seeps upwards over the surface of the brain to be absorbed back into the bloodstream and also down into the spinal canal, where samples can be drawn off for testing when a lumbar puncture is performed. A tiny central canal traverses the length of the spinal cord, but this is merely a cul-de-sac.

The fluid escapes from the fourth ventricle by means of apertures which are well-known eponyms—the foramina of Luschka and Magendie. The foramina described by Hubert von Luschka are on each side, while the one in the middle is named after François Magendie, the great French neurologist. Luschka was a professor of anatomy in Tübingen, Germany and both men described their escape holes in the same year, 1855. The foramina are important and any congenital deformity or later infection which blocks them will cause the ventricular spaces to swell, leading to the distressing condition of hydrocephalus, or water on the brain.

The brain stem is the axis of the brain which continues upwards from the spinal cord. The oldest part of the brain in evolutionary terms, the brain stem is complex. Its main parts are the peduncles, which are the roots of the cerebrum and cerebellum, the midbrain, the pons, and the medulla, which merges into the spinal cord.

Peduncles mean feet and they join the thick

François Magendie introduced the methods of experimental physiology into pathology and pharmacology. Born in Bordeaux in 1783, he studied medicine at the College de France where, in 1831, he was appointed to the chair of medicine. Claude Bernard was one of the many whom he taught and inspired. His work extended to many fields, including cardiac function, digestion, and the effect of drugs and poisons on the body. Magendie, who founded the first journal for experimental physiology, rejected the idea of an external unified life force, believing that the function of organs in a body give life to the whole. He died in 1855.

midbrain and the pons. Pons is Latin for a bridge, an appropriate name for the point where connecting nerve fibres cross over from one side of the cerebellum to the other. The pons is still called the pons Varolii, after Constanzio Varolio, a professor of anatomy and surgery at Bologna and later at Rome, who described it in his *Anatomy* published in 1573.

The medulla oblongata is a most important part of the brain, for it contains some of the vital centres controlling breathing and heart rate.

Within the brain stem is a scattered collection of grey and white matter which is known as Dieters' formation, commemorating the same Otto Dieters as is remembered in Dieters' nucleus. Dieters' formation is part of what is now known as the reticular formation, a system with an uncertain function.

The spinal cord is really a long extension of the brain, rather than a separate structure. The two together form the central nervous system, or CNS as it is universally known to doctors and and students. Both parts are encased in a strong protective shell of bone, sheathed in the dura mater, and buffered by the cerebro-spinal fluid.

The spinal cord is considerably shorter than the spine itself. It stops at the level of the small of the back, although the dural sheath continues for some distance, containing the leash of nerve roots which is called the cauda equina, the horse's tail. From the tip of the spinal cord a long, thin strand emerges, to anchor the cord to the lower end of the spine. This is the filum terminale.

On the back aspect of the upper spinal cord are two bulges which carry a large collection of nerve fibres. Known now as the fasiculus gracilis and fasiculus cuneatus—the slender bundle and the wedge-like bundle—they were formerly called the columns of Goll and Burdach.

Rudolf Virchow, one of the world's greatest pathologists, laid to rest the ancient theories of disease. Born in Germany in 1821, he read medicine at Berlin. Early in his career he began to study the connection between cells and pathology, identifying the cells, rather than an organ or system, as the site of disease. Concerned with problems in public health, especially epidemiology, Virchow was deeply involved in national politics, concerning himself largely with improving general health services. One of his hobbies was anthropology, and in 1869 he helped found the German Anthropology Society. Virchow died in 1902.

Giulio Casserius was born in 1561 in Piacenza in Italy and went to Padua for his medical studies. There he became Fabricius's assistant and successor and was one of William Harvey's teachers. Casserius's own interest lay in the organs of hearing and of speech. In 1601, his book on the comparative anatomy of the ear was published in Venice. The book was illustrated with copperplate engravings, notable because the anatomical detail was combined with artistic beauty. Casserius lived until 1616. His work was an important contribution to the development of new techniques of anatomical study by means of accurate observation.

Friedrich Goll was a Swiss anatomist who studied with some of the most illustrious men in Europe during the mid-nineteenth century, including Claude Bernard, Virchow, Ludwig, and Kölliker. Goll, when professor of anatomy at Zürich, described the fasiculus gracilis in 1860. About fifty years earlier, Karl Burdach, a German anatomist and physiologist, described the posterior column, which he designated fasiculus cuneatus.

Clark's column in the spinal cord is really an elongated nucleus rather than a fibre tract. It commemorates the English anatomist Jacob Lockhart Clarke who described it in 1851.

An extensive part of the nervous system lies outside the cerebro-spinal zone and many of these structures commemorate ancient as well as relatively modern men of medicine.

Casserio's perforating nerve, for example, is the musculo-cutaneous nerve which supplies the muscles and outer side of the upper arm. Giulio Casserio, known as Casserius, was the successor to Fabricius in the famous anatomy chair at Padua. He described this nerve in his *Anatomical Tables* which was published in 1627, eleven years after his death.

Ganglia, or knots in Greek, are collections of nerve cell bodies which are external nuclei for sensory and autonomic nerves. A number have specific names; some, such as the Gasserian ganglion on the trigeminal nerve of the head, are eponymous.

Johann Gasser was a professor of anatomy at Vienna from 1757 until 1765, and he died while still a young man. The sensory ganglion was named for him by one of his students, Raymond Hirsch, in 1765. It has been suggested, however, that this doubtful tribute to a little known anatomist may actually be the result of an error in the spelling of the name Casserius and that Casserian ganglion was intended.

The trigeminal nerve is the largest of the twelve cranial nerves that arise directly from the brain and it gets its name from the three prominent branches that come off the Gasserian ganglion. As well as being the motor nerve for the jaw muscles, this nerve receives sensation from much of the surface of the head. The large ganglion which contains the sensory nerve cells lies in a recess in the base of the skull called the cave of Meckel after Johann Friedrich Meckel, an eighteenth-century Berlin professor of anatomy.

After the trigeminal, the next most important nerve serving the face and head is the facial nerve, the fifth cranial. This motor nerve of the face was first described by one of the leading British anatomists of the nervous system, Sir Charles Bell. He also described the effect of paralysis of this nerve which became known as Bell's palsy. (The canal in which the facial nerve runs through the temporal bone is the Fallopian canal, named after the revered Gabriele Fallopio.) Sir Charles Bell is also remembered in the long thoracic nerve known as Bell's nerve, which supplies the region of the armpit and the side of the chest, which he described in 1829.

In the nerves themselves, some indelible eponyms have survived the ruthless pruning to revise anatomical nomenclature. Around most of the mature nerves, a layer of a fatty substance called myelin coats the axon, or actual conducting fibre, rather like the insulation of an electric cable coats the inner metal conductor. This myelin is confined within a delicate sheath, as a sausage is packed within its skin. The sheath is called the neurilemma, or sheath of Schwann, and is formed from so-called Schwann cells which have a flattened nucleus buried under the neurilemma.

Each nerve fibre resembles a piece of bamboo, with abrupt nodes spaced at intervals between long, straight segments, again rather like a string of thin, elongated sausages. At each node the myelin sheath is sharply constricted and one Schwann cell occupies the interval between two nodes. The constriction is known as the node of Ranvier.

In 1839, Theodor Schwann, a German anatomist and a pioneer of neuro-anatomy, described the cells which bear his name. Louis Antoine Ranvier, a French histologist, wrote on the nodes of the nerve sheath in 1878.

At the ends of the sensory nerves, various end-organs are responsible for initiating the impulses that are sent up the nerves to the spinal cord and brain. The detection of different types of stimuli require different nerve endings. Light touch, pressure, temperature, pain, and awareness of muscle tension and joint position are some of the various sensations that are picked up by the different structures that lie at the business end of various nerve fibres. Most of these have names associated with them.

Meissner's corpuscles, tiny cylindrical organs which detect touch, are found in profusion in the skin, especially of the fingers and lips. Georg Meissner, a German anatomist and physiologist, described this organ, which he investigated with Rudolf Wagner at Göttingen in 1853.

Another end-detector is the Pacinian body, a larger ovoid corpuscle with many layers like an onion, which picks up pressure and vibration. The name is taken from that of Filippo Pacini, an Italian anatomist, but the corpuscles were first described in 1717 by Abraham Vater, the same Vater whose name is attached to the opening of the bile duct. Pacini, who got the eponymous credit, was a professor of anatomy and physiology at Pisa, and published his description of the pressure receptors in 1840.

Krause's end-bulbs are a type of end-organ whose function is not clear. Their eponymous origin is also uncertain since there were two Krauses, father and son. The first was Karl, a professor at Hanover, and the son was Wilhelm, professor at Göttingen and Berlin, who has a whole galaxy of eponyms to his credit.

Another nerve ending found in the skin, genitals, muscles, and tendons is the Golgi-Mazzoni body. The structure is named after the Italian anatomist Camillo Golgi, but the Mazzoni component is obscure since little is known of Vittorio Mazzoni, save that he published his findings in a Bolognese medical journal in 1891. Also named after Camillo Golgi, however, are Golgi corpuscles, end-organs of a fusiform shape which are found at the junction of muscle and tendons.

In the sensitive skin of the fingers, Ruffini bodies are to be found; these nerve endings were described in 1898 by another Italian anatomist, Angelo Ruffini.

The Italians seem to have had a heavy stake in the end-organs.

Charles Bell was born in 1774 in Edinburgh. The leading British anatomist of his time, he worked principally in the field of neurology. He received his medical training at Edinburgh where, in 1836, six years before his death, he became professor of surgery. He wrote a number of important works. *A New Idea of the Anatomy of the Brain and Nervous System*, published in 1811, had been called the Magna Carta of neurology. In it Bell reported the experimental demonstration of the motor function of the anterior roots of the spinal nerves. In 1829, Bell received the first medal awarded by the Royal Society and was knighted two years later.

Theodor Schwann was the first to recognize that the cell is the basic unit of all living things. Born in Rhenish Prussia in 1810, he went to Berlin where he worked with Johannes Müller. He taught anatomy and physiology at Berlin, at Louvain, and, later, at Liège. Schwann investigated the histology of the nervous system and was also interested in such natural processes as fermentation. He isolated the enzyme pepsin and coined the term metabolism to define the chemical changes in living tissue. Schwann lived to be seventy-two, but it was during the four years spent under Müller's influence that most of his best work was done.

Vital to the understanding of the central nervous system is the concept of the reflex action, when a movement is initiated by an autonomic pathway from sensory nerve, through the spinal cord, and back through a motor nerve to a muscle, without the participation of the brain. This mechanism was hinted at by René Descartes in his book *De Homine*, which was published in 1662, twelve years after his death. And Pavlov, more than two hundred years later, acknowledged Descartes's work as the stimulation for his own research.

Descartes was a ventricle man, in the medieval mould. He believed that spirits passed down the nerves and it was he who thought that the pineal, the tiny, pea-sized gland deep at the back of the brain fissures, was the seat of the soul and controlled all brain function by modulating the flow of spirits in the ventricles. He was, however, mechanistic in outlook and related all functions to a mechanical model. His glimmerings of reflex action came from his description of the way a muscle movement arises from a visual image. Descartes alleged, for example, that light from an object formed an image on the retina of the eye, which was connected by the optic nerves to the walls of the brain ventricles. This made animal spirits travel to the pineal, which caused more spirits to travel down the nerve to the arm to initiate movement. In this scheme, the brain's grey and white matter played no more than a structural, supporting role.

Thomas Willis, the physician who was one of the dominating figures in English medicine in the seventeenth century, had the opposite conviction. He believed, correctly, that it is the solid matter that functions, not the fluid-filled ventricles. He was not the first, however, to suggest the brain substance as the important site of function. François de la Bois, the French anatomist who was known as Sylvius, had suggested only a few years earlier that the animal spirits came from the cerebral cortex.

In 1664, Thomas Willis published his book on the brain, which he had written in collaboration with Richard Lower. Some of the drawings included in Willis's book *Cerebri Anatome* were executed by Christopher Wren. There is reason to believe that the famous circle of Willis was suggested by this artist-architect who is remembered far more for his restoration of St. Paul's Cathedral in the city of London, than for his anatomical understanding.

With some circumspect sidestepping, to avoid offending the theologians who were casting suspicious eyes at these probings of the brain and, they believed, the soul, Willis took the old three-cell theory and rerouted the functions. He declared that all incoming sensations were received in the large basal ganglia, the mass of grey matter called the corpus striatum that lies in the depths of the brain alongside the main ventricles. He stated that the white matter of the cerebral hemisphere held the power of imagination, and that the grey matter of the cerebral cortex was the seat of memory.

Willis further wrote (and was not far from the truth) that while voluntary muscle activity was initiated in the cerebral cortex, more vital functions were controlled by the cerebellum. This localization theory was eventually disproved, but it was nevertheless a landmark in the laying of the ghosts of medieval fallacies.

In the same decade as Willis's book was published, nerve endings were seen by the indefatigable microscopist Marcello Malpighi. Experimental work on animals became more exact at that time, with decerebrate and decapitated frogs available for the investigation of the all-important reflex actions.

In the early years of the eighteenth century, memories of Galen still lingered and the idea of reflex action was still not current. In 1751, Robert Whytt, a Scottish physician and one of the outstanding professors of the Edinburgh of his day, described the reactions of animals who had certain parts of their brains and spinal cords removed, showing that the spinal cord is essential for reflex movements. He proved, too, that reflex movements might not be perceived by the individual, since they are performed without the involvement of the higher brain centres.

There was great interest in the eighteenth century in animal electricity, which included studying such creatures as torpedoes and electric eels that produce quite high voltages as well as the effect of small current on isolated frog muscles. Luigi Galvani, a professor of anatomy at Bologna, is well known for this work. The electrical nature of the nerve impulse began to

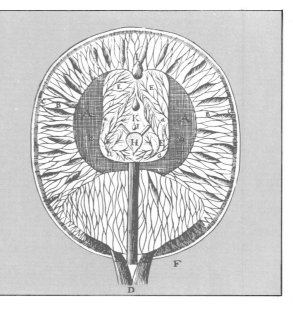

René Descartes had many original concepts about the structure and function of the brain which were, unfortunately, based on speculation. He believed that the pineal gland, H in the illustration right, exerted control over brain function by mediating the flow of spirits to and from the ventricles. Many of the illustrations in his book *De Homine* were diagrammatic. The drawing left, for example, is the simplest representation of his concept of brain function and is curiously reminiscent of medieval schematic drawings. Descartes's work, however, particularly his ideas of the reflex, did raise valuable questions about nervous function.

be appreciated and Galen's hydraulic theory faded away.

In the nineteenth century, research in neurophysiology really took off, as it did in all other branches of medical science. DuBois Reymond, a pupil of Johannes Müller, the distinguished German professor of anatomy, refined the crude electrical measuring instruments of the day sufficiently to detect electrical currents in an animal nerve. In a memorable experiment in 1850, he demonstrated for the first time that the nerve impulse is an electrical wave passing along a nerve trunk. Hermann von Helmholtz, another pupil of Müller, succeeded in the same year in measuring the speed of transmission of the nerve impulse—between twenty-five and forty meters per second, much slower than had been guessed.

Much of this work in the area of electricity was solely related to peripheral nerve fibres and the basic mystery of the brain was not being significantly advanced, although its detailed anatomy was slowly being unfolded.

The idea planted by Willis that different areas of the brain have specific functions was looked into again and in 1811 Jean Legallois, a French surgeon, discovered that the medulla oblongata is responsible for the control of breathing. This discovery was made, as were so many other neurological findings, by observing the abnormality resulting from part of the brain being damaged.

Brain damage has proved an indispensable tool in the investigation of brain function. The experimental surgical damage inflicted on animals, as well as the spontaneous damage seen in patients as the result of disease and injury, created the opportunity to localize the origins and pathways of nerve impulses in the nervous system. One whole area of neuro-anatomy, for example, involved tracing the columns in the spinal cord and brain stem where certain bundles of nerve fibres, or tracts, pass up and down and often cross wholly or partially from one side to another. It would be virtually impossible to dissect them out from among the hundreds of thousands of almost parallel fibres in adjacent tracts. Consequently, the use of damage was invoked to detect their position. Dead tracts stain differently from healthy fibres—indeed, may sometimes be seen without staining. When a tract was killed because it was cut off from the cell bodies in the corresponding nucleus, then the degenerating fibres could be picked out after a lapse of time and their topography mapped out.

Legallois noted the respiratory problems in a person with a damaged medulla. Many other functions were traced by correlating the disease symptoms with the physical damage in the brain, usually at autopsy. The man with left-sided paralysis was found to have had a cerebral

Christopher Wren, the architect of St. Paul's Cathedral, translated science into visual forms. Born in Wiltshire in 1632, he was tutored by his brother-in-law and at the age of fifteen he designed anatomical models and experiments. Wren took his degree in astronomy at Oxford. In 1657, he became professor of astronomy at Gresham College in London and joined the Invisible College, the group of polymaths who later founded the Royal Society. He worked with Robert Hooke and Robert Boyle in physiology and with Thomas Willis in anatomy. He was "frequently present at our dissections . . . so that the work could be more exact," and his elegant rendering of the circle of Willis is the most famous drawing of the brain of the time. Wren became fascinated with architecture. He travelled to the continent and met the famous Italian Bernini in Paris. After the Great Fire of 1666 destroyed much of London, Wren's designs were adopted for the rebuilding of St. Paul's. Until his death at the age of ninety-one, Wren worked to raise architecture to an art that was "capable of beneficial scientific enquiry."

haemorrhage on the right side of the brain. The woman with the weak right leg was seen to have old softening in the left motor pathway. People who burned their fingers because they could not feel temperature changes were found to have degeneration in the centre of the upper spinal cord—a disease known as syringo-myelia, meaning hollow cord. Slowly, the relationship between loss of movement and physical damage created a picture of localization of function.

Sometimes, the method failed. The most famous case was that of Phineas Gage, a quarryman in the United States, whose head was completely transversed by a three-foot crowbar during an explosion in 1848. Although a doctor could put fingers inside Gage's skull and insensitive brain, the man recovered in minutes and lived for another twelve years. His only disability was a change in personality for he was an

Luigi Galvani was the Father of Electrophysiology and with his experiments he provided the stimulus that ushered in the age of electricity. Born in 1737 in Bologna, like his father he became a doctor. He was lecturer in anatomy and professor of obstetrics at Bologna, but was more interested in experimenting with a machine that produced electricity. His idea that animal electricity was a new kind of force gave rise to the term galvanism. When Napoleon invaded Italy, Galvani refused to swear allegiance to the new republic. He was dismissed from his posts and denied his salary. He was soon recalled, but died shortly afterwards in 1798.

Hermann von Helmholtz had the remarkable ability to coordinate insights from all fields of study. Born in Potsdam in 1821, he was a frail child, and was educated at home by his father who taught him Latin, Greek, Hebrew, Arabic, English, French, and Italian. He studied medicine in Berlin with Johannes Müller and became his assistant, but he rejected his teacher's theory of "vital forces." Von Helmholtz taught at Königsberg, Bonn, and Heidelberg and gradually his interests evolved from physiology to physics. In 1888, he became the first director of the Physico-Technical Institute in Berlin, where he worked until his death in 1894.

Two years later, Marie Jean Pierre Flourens, the French anatomist, was able to prove that various parts of a dog's brain perform specific functions by the now familiar method of cutting connecting tracts and fibres in the brain and observing the progressive deterioration of functional ability. He demonstrated, by removing portions of the brain in pigeons, that the cerebrum is the organ of thought and will power and that the cerebellum is concerned with the coordination of muscular movements.

Later in the century, John Hughlings Jackson, a doctor at the National Hospital for Nervous Diseases in London, described various defects that brain-damaged persons suffer and stated that there must be a specific area controlling motor activity. It was this man whose name is commemorated in Jacksonian epilepsy.

In 1861, Pierre Paul Broca, then a professor of clinical surgery in Paris, published his findings, resulting from his study of patients with speech defects, that the speech area is localized on one side of the brain, usually the left side in right-handed people. Within the next ten years, Karl Wernicke, the German neurologist, confirmed Broca's work and described another region where sound is received and appreciated. These two parts of the brain's cerebral cortex have since been universally known as Broca's area and Wernicke's area.

As technological advances were made a new method of localizing function became available. In the European wars late in the nineteenth century, battle victims sometimes had parts of their skulls blown away, so that the brain surface was exposed. Eduard Hitzig, a German military surgeon, took the opportunity to apply wires from a small battery to the cortex and observe what happened. As a result of these ghoulish experiments he discovered that eye movements and other reactions could be stimulated.

Hitzig later performed animal experiments of the same nature with a colleague, Gustave Fritsch. (For lack of other facilities some of this work was performed on the dressing table of Fritsch's wife.) Their investigations proved that repeatable and constant movements could be obtained from some areas of the brain, while others seemed unresponsive.

The use of such electrical stimulation of the brain (ESB) became widespread later in the century for investigating localization of function. Sir David Ferrier, an English neurologist, published the most exact maps of cortical motor activity. He and others produced minutely accurate details of the excitability of various convolutions in the brains of animals and correlated much of this with known disease and injury damage in human patients. This ESB work had limitations, however, because some unresponsive parts of the brain remained stub-

example of an inadvertent frontal leucotomy. The frontal lobes perform no motor or sensory function, but in the nineteenth century, when doctors were still groping for true understanding of tangible localization of nervous activity, such phenomena were inexplicable.

The concept of strict localization of function was outlined in 1811 by Sir Charles Bell, the English anatomist, as a counter argument against the still-lingering idea of the *sensorium commune*, which held that the brain functioned as a whole unit without any particular zones for specific functions. Charles Bell rather blotted his reputation, however, by antedating a paper in 1821 on work which was published the following year by François Magendie on tracing motor and sensory pathways.

bornly obscure until sufficiently sensitive electrical instruments were developed which could detect minute currents in the brain tissue, necessary to investigate the sensory areas of the brain. Whereas ESB sent electricity into the brain to cause nerve impulses to pass outwards to produce muscle movements, the sensory input from the periphery of the body to the brain could only be detected by picking up the tiny electrical messages that reached the sensory areas of the cortex.

It was the 1860s before any galvanometer was good enough to register a millionth of an ampere and this was first used to repeat DuBois Reymond's work on nerve fibres, where the electrical changes were greatest.

The galvanometer was first used on the brain by Richard Caton in Liverpool. In 1875, he reported that he had detected electrical activity in the exposed brains of rabbits and monkeys. It was a diffuse, confused current variation, but changed when an animal performed some specific act like moving or eating. Caton was also able to pick up currents in the optical area of the brain cortex when a bright light was shone into the eye. This was the first example of the detection of sensory input to the brain.

Even more important, Caton tried putting the wires of his galvanometer on the skull and scalp, instead of directly on the brain—and he still picked up the currents. This opened the door to the highly significant application to humans, who might not appreciate having their brains exposed to satisfy the curiosity of researchers. This discovery was the birth of the electroencephalograph (EEG), which was to be developed later into a most important clinical tool for neurologists and neurosurgeons.

Adolph Beck, a German, greatly extended Caton's detailed investigation of brain electrical activity. In his doctoral thesis of 1891, he related how the brain had a continuous low-level activity, which was suddenly peaked at certain points when some sensory stimulus came in. After Beck published his work in a medical journal, a professor from Vienna named von Marxow revealed that he had done similar research in 1883 and produced a sealed letter held in the vaults of the Vienna Academy of Science to prove it. He also had detected electrical activity through the scalp. Beck and von Marxow began disputing about who did what first, but the argument was promptly settled by a letter from Richard Caton in England, who had done it all in 1870. Then, a Russian, Danilevsky, wrote to the editor to say that he had discovered electrical activity in the brain in 1876 and published the following year. Caton, however, still led the field, but the Russian proved himself ahead of the German and the Austrian.

The other great advance of the nineteenth

Santiago Ramon y Cajal was born at Petilla da Aragon in Spain in 1852. He became interested in the microscope when he was twenty-five, and from that point he devoted his energy to advancing microscopy techniques. He amazed the scientific world in 1889 when he presented the results of his work on the nervous system to the German Anatomical Society. As resistant as the world had been to his findings, in the years following that meeting he received many honours, culminating in the Nobel Prize for medicine in 1906. In 1920, King Alfonso XIII of Spain founded the Institute Cajal at Madrid where Cajal worked and taught until his death in 1934.

century was the neurone theory, which was the anatomical basis of all understanding of the nervous system. When better microscopes and developments in staining tissues for microscopic examination came along, new knowledge of the minute structure of nervous tissue made many functional matters clear.

In 1889, a Spanish histologist, Santiago Ramon y Cajal, demonstrated a new method of staining nerve cells which revolutionized the examination of brain and nerve tissue. For his exquisite staining techniques on the cells of the cerebral cortex, Cajal shared the 1906 Nobel Prize in medicine with Camillo Golgi. Cajal's name is still remembered in certain of these nerve cells—those which have several branches orientated in a horizontal direction and were described by him at the beginning of the twentieth century. But before this, others had advanced the study of the neurone, or nerve cell.

At the end of the eighteenth century, microscopists had alleged that nerve fibres were hollow tubes, or tubes filled with gelatinous fluid or strings of globules. Their poor instruments would not resolve detail well enough, but in the early part of the nineteenth century, the invention of the achromatic lens permitted Johannes Purkinje, the Bohemian physiologist and professor at Breslau, to make a much better job of describing the nerve cells. He showed that the cell had a body and a nucleus, together with thin tentacles sticking out from the edges, one of which was much longer than the others. His name remains attached to the most spectacular of all nerve cells, the large, flask-shaped cells of the cerebellum now known as Purkinje's cells.

It was soon realized that the long processes seen by Purkinje were the actual nerve fibres and that every fibre in the body had to have a parent cell body, even though this might be a few feet distant, as in the nerves of the fingers or toes. Much effort was expended in inventing better ways of staining nerve cells and fibres.

The interior of the nerve cell was investigated later in the nineteenth century by Franz Nissl, a German neurologist. Using a stain of basic dyes, he was able, in 1884, to discern particles in the cytoplasm of nerve cells. Nissl granules, or bodies, have become an integral part of the vocabulary of neurohistology.

Camillo Calleja, a professor of anatomy at Madrid, published an account in 1893 of certain cells in the cortex concerned with the sense of smell. Known as Calleja's islands, these cells are in the hippocampal gyrus on the undersurface of the brain.

As part of the tremendous interest in microscopic anatomy and pathology that flourished in the nineteenth century, the Italian Camillo Golgi, whose name has been perpetuated in the Golgi apparatus inside almost all cells, developed a silver stain for nerve tissues and shared in 1906 the Nobel Prize with Cajal, who brought this work to a peak of refinement.

As a result of the discoveries of these micro-anatomists, a clearer idea of nerve function was gained—and in the nineteenth century function largely meant the reflex action.

The reflex is a movement brought about by a sensory impulse, without the participation of the higher brain centres. It functions either through the spinal cord or through the cranial nerve and brain stem which, for this purpose, can be thought of as an upper prolongation of the spinal cord. When a bright light is shone in the eyes, for example, the pupils constrict—but this does not have to be voluntarily willed to happen, just as a person does not have to remember to keep breathing or to snatch a hand away from a hot saucepan.

The nature of the all-important reflex, without which life would be impossible, was a matter of great interest in the nineteenth century. In 1811, Sir Charles Bell found that of the two nerve roots that emerge from each level of the spinal cord on each side, the front one is motor in function, that is, it initiates muscle action, and the rear one, he said, is non-motor. In 1821, however, his pupil John Snow, and in 1822, François Magendie, proved that the rear or posterior roots are sensory, bringing messages in from the body and its environment. This became known as the Bell-Magendie law.

Marshall Hall, a Nottingham physician, is sometimes called the Father of the Reflex and his name is certainly indelibly linked with this action. Although he seemed unaware of similar work done previously, in 1833 Hall wrote up a mass of experimental work on animals, especially chickens, in which he demonstrated reflexes in headless and bodiless birds in all kinds of situations. Hall proposed the concept of the spinal brain, which correctly supposed that the spinal cord is more than a mere thick cable of nerves going up and down from the skull, but has its own initiating role. It is, as he said "superior to the real brain, for the spinal one is always awake and alert while the brain needs sleep."

Hall's views aroused much antagonism, because it was accepted that the soul resided in the brain and it smacked of heresy to set up a rival brain in the spinal cord. Hall was banned from publishing in the journal of the Royal Society because of the controversy, but finally his views were accepted and in 1850 he gave the prestigious Croonian Lectures on the subject.

Perhaps the best-known name in neurology in the nineteenth century was that of Charles Sherrington. He pursued the same work on reflexes, which carried over into the twentieth century. Sherrington began with a painstaking study of the nerve pathways in the spinal cord and brain stem. This culminated in a lecture course at Yale University in 1906, a published version of which is one of the classics of neurology, and is still being reprinted.

Sherrington worked out the reciprocal nerve supply of antagonistic muscles, whereby one muscle relaxed when another contracted. He described decerebrate rigidity, a state which afflicts an animal whose brain stem has been divided at a certain level and which also occurs in an injured or diseased human patient. The inhibiting effects of the higher brain centres is lost in such cases, so that the reflex brains in the cord have unchecked power to keep the muscles in a fixed, characteristic posture.

Using decerebrate animals, Sherrington invented an electrical flea, a pin with a weak current to simulate an irritant sensory stimulus on the animal's skin. Using this, he showed that the reflex was a highly complex and integrated response, far different from the "one nerve in—one nerve out" concept of early investigators. In his ninety-one years, Sherrington added a new dimension to neurophysiology, as well as to standards of scientific writing; his perfect command of language and his humanist outlook made his papers a literary as well as a medical delight.

Sherrington was mainly concerned with spinal and brain-stem reflexes. Others investigated higher reflexes, mediated through the brain. One of the first to write on this subject, Ivan Sechenov, got into trouble with Russian authorities in 1866 because they considered this materialistic view of the mind inconsistent with Christian and moral doctrines. The charge was eventually dropped. Sechenov, after whom the main medical college in Moscow is now named, was a brilliant physiologist who applied the general principles of reflexes to higher brain function. He believed that the central arc of the three-part reflex system was vital in these higher functions and could be elaborated and shunted about within the deeper parts of the brain and up to the cerebral cortex to provide the basis for all intellectual functions.

His hypothesis went far beyond his experimental base and beyond his own time, but his controversial writings were a great stimulus to further research. It has been said that Sechenov was the model for Bazarov in Turgenev's novel *Fathers and Sons*. Many discoveries in the nervous system, attributed to others, including Sherrington, were actually made by Sechenov, and he was without doubt the pioneer and, perhaps, the greatest exponent of Russian neurophysiology.

Sechenov was the direct inspiration for Pavlov, whom he never met, but Ivan Pavlov always recorded his debt to the great man. Pavlov applied the reflex concept to the highest levels of nervous function, in that he showed that purely psychic stimuli could evoke such marked physical responses as the sight of food causing salivation and the outpouring of stomach juices. He was awarded the Nobel Prize in 1904 not for his work on conditioned reflexes, but for research into digestion, although it was his work on reflexes that brought him world-wide fame.

In the twentieth century, the progress of research in neurophysiology and in the ultrastructure of neuro-anatomy leaped forward. In 1939, the first electron microscope became available and its enormous magnifying power and resolving ability opened new windows into the understanding of the structure of the neurone. Twenty-seven years later, the scanning electron microscope allowed three-dimensional views to be seen and these advances clarified much that was ill-understood, especially concerning the synapses, the junctions between nerve cells where the impulses have to "jump the gap."

The nature of this jump has also become clear, in that it is a chemical rather than an electrical phenomenon. The passage of the nerve impulse down the nerve fibre is electrical; a wave of depolarization (a change in the electrical potential of the surface compared to the centre) runs along the nerve, at a speed which varies according to the size of the fibres. This depolarization has been found to require rapid changes in sodium and potassium for its production and for the restoration of the fibre to a state permitting another impulse to pass. But when the electrical component reaches the end, the stimulation of the adjacent cell is made by chemical means. The chemicals are different for ordinary nerves and for sympathetic nerves.

Those concerned with the unravelling of this extremely complex process include Claude Bernard who, when working on the effects of curare, a South American arrow poison, found

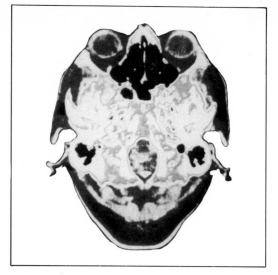

Thermography is a technique for finding and measuring variations in temperature throughout the body tissues, which may be indicative of hidden abnormality or disease. A unique way to view the topography of such complex organs as the brain, the thermograph—used in hospitals as a diagnostic aid—resolves as a colour-coded X-ray film.

that it blocked certain types of nerve transmission. Otto Loewi, an Austrian, made great strides towards substantiating this theory when, in 1921, he extracted some substance which had an effect on frogs' hearts. He realized that there were two different substances and found that one was adrenaline. In 1929, the English biologist Henry Dale named acetylcholine as the other nerve mediator. These discoveries had enormous implications for therapeutic medicine, as well as for physiology. A direct result of Claude Bernard's work, for example, has led to the use of curare-like substances in anaesthesia, to relax muscles during operations, allowing surgeons better access.

All branches of anatomical and physiological science, as well as pure physics and pharmacology, have played a part in the elucidation of the structure and function of the nervous system. The basic question of the nature of consciousness and "self"—what some would have considered the location of the soul—still eludes us. But places in the brain where the soul can hide are getting fewer and smaller. It is certainly not in the pineal, as was once thought by Descartes. And many medical men now have the uncomfortable feeling that when the last hiding place is opened up, there will be nothing inside. This is not to say that the psyche does not exist, but that it is a function of the whole brain—and maybe even including the spinal cord—functioning as an integrated totality.

Recent work on micro-chip computers has shown that when a certain threshold of complexity is reached the computer begins to have strange properties that are beyond the bounds of theoretical anticipation. The total capacity becomes greater than the sum of the individual parts—a state which the human brain seems to have reached several million years before micro chips were contemplated.

The Ears

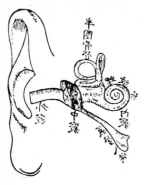

This accurate diagram of the ear is a woodcut which appeared in a nineteenth-century Chinese translation of *Compendium of Anatomy* by Benjamin Hobson, a British missionary.

The ancients had peculiar ideas about the ears. In Babylon the ears were believed to contain the human will because they received commands and were thought to be responsible for carrying them out. In an Egyptian medical papyrus, written in the second millennium B.C., it was stated that the ears were the organs of respiration as well as of hearing and that "the breath of life enters by the right ear and the breath of death by the left ear." In the fifth century B.C., the Greek physician Alcmaeon of Croton described the auditory tubes named centuries later after Eustachius, but, nevertheless, he seemed to be influenced by the Egyptians for he declared that goats breathe through their ears.

Hippocrates believed that sound was conducted directly to the brain from the ears through the bone of the skull. Galen, perhaps fortunately, had little to say about hearing, except that sound was due to vibrations in the air, a perceptive observation at so early a date.

Early anatomists, contemplating the visible

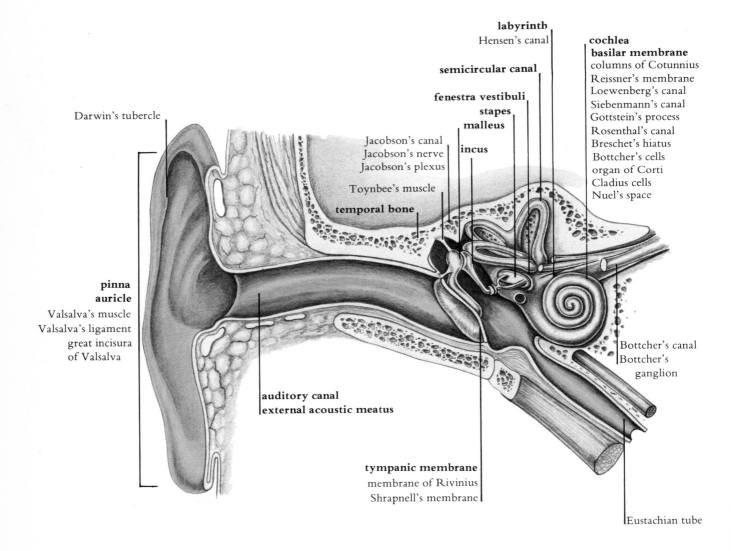

labyrinth
Hensen's canal

semicircular canal

fenestra vestibuli
stapes
malleus

incus

cochlea
basilar membrane
columns of Cotunnius
Reissner's membrane
Loewenberg's canal
Siebenmann's canal
Gottstein's process
Rosenthal's canal
Breschet's hiatus
Bottcher's cells
organ of Corti
Cladius cells
Nuel's space

Darwin's tubercle

Jacobson's canal
Jacobson's nerve
Jacobson's plexus

Toynbee's muscle

temporal bone

pinna
auricle
Valsalva's muscle
Valsalva's ligament
great incisura
of Valsalva

auditory canal
external acoustic meatus

Bottcher's canal
Bottcher's ganglion

tympanic membrane
membrane of Rivinius
Shrapnell's membrane

Eustachian tube

ear, or auricle, could have had no clue as to the infinite complexity of the human ear and to its dual role as the organ both of hearing and of balance. For the ear, a delicate assembly of mechanical and nervous structures, has two inner parts, both housed inside the same block of dense bone in the base of the skull.

The hearing function is performed by a membranous drum stretched across the inner end of the external auditory meatus, or ear hole. Sound waves are directed along this auditory channel, causing the eardrum, or tympanic membrane, to vibrate.

Behind the eardrum is the cavity of the middle ear. Here the vibrations are magnified mechanically by a trio of miniaturized bony levers called the ossicles, little bones. The smallest bones in the body, the ossicles are named for their shapes. They are called the malleus, the incus, and the stapes—the hammer, the anvil, and the stirrup. The malleus is attached to the inside of the eardrum and via the incus it moves the stapes bone when the drum vibrates. The base of the stirrup acts as a tiny plunger in an oval hole in the bony wall of the internal ear called the fenestra vestibuli, the window of the vestibule.

From the middle ear, sound waves are transmitted to the spiral, fluid-filled cochlea of the inner ear, which is deep in the petrous, or stony, part of the temporal bone of the skull.

Pythagoras, Alcmaeon's teacher at Croton, made no direct contribution to the understanding of the ear. His pioneering genius in mathematics and physics led him to show, however, that the pitch of a note produced by a stretched string depends upon its length. This fundamental truth explains the construction of the cochlea. The cochlea is a spiral with decreasing diameter, and accommodates the basilar membrane, which is covered with nerve receptors for sound—the longer ones detect lower pitch, the shorter ones high-frequency sounds. From here nerve impulses travel to the brain by way of the auditory nerve.

The middle ear is connected to the back of the throat by the Eustachian tube, which is named after the great anatomist of Rome Bartolomeo Eustachio who described it in 1562. One of the best-known eponyms in the body, the Eustachian tube's function is to equalize the air pressure in the middle ear with that of the atmosphere, otherwise sudden changes in pressure, besides interfering with hearing, would cause the eardrum to burst. It is for this reason that chewing gum or hard candies are handed out on unpressurized aircraft, so that the action of swallowing will help to pass air up or down the Eustachian tubes.

The mechanisms for maintaining balance and posture are housed in the inner ear. Here the three semicircular canals, set at right angles to each other, detect the slightest movement of the head in any particular plane. Swirling of the fluid in the canals triggers nerve impulses related to balance. Also involved in posture and balance are two fluid-filled chambers, the utricle and saccule, which keep the brain informed of the body's orientation in space. Nerve impulses from these chambers and from the semicircular canals pass along the vestibular branch of the auditory nerve to centres in the brain which govern balance and physical coordination.

Knowledge of the anatomy of the ear began to accumulate in the middle of the sixteenth century when several of the great anatomists took an interest in the organ of hearing. Gabriele Fallopius, who succeeded Realdus Columbus as professor of anatomy at Padua, introduced the two most important terms in relation to the functions of hearing and balance. These were the cochlea, from the Greek word for snail, because of its shape, and the labyrinth for the inner ear, which he named after the maze of tunnels where the mythological Cretan minotaur lived. Fallopius is commemorated in the ear in the aqueduct of Fallopius, a canal in the bone which passes through the tympanic cavity, carrying the nerve known as the chorda tympani.

In the middle of the sixteenth century, Giovanni Filippo Ingrassias, the Sicilian Hippocrates, was the first to describe the stapes, the innermost of the tiny ossicles. A Dutchman, Volcher Coiter, who was a student of Fallopius at Padua and of Eustachius at Rome, had settled in Nuremberg and wrote on comparative anatomy in

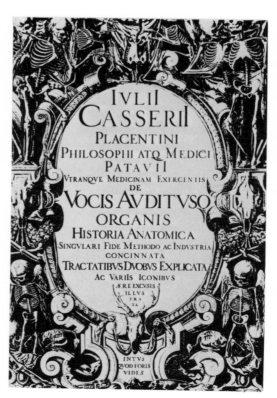

Giulio Casserio, one of William Harvey's teachers at Padua, made a special study of the ear and larynx. In his beautifully illustrated book *De Vocis Auditusque Organis,* published in 1601, he gave an account of the ligaments of the delicate malleus.

This illustration of the human ear was drawn by Eustachius and appeared in his *Opuscula Anatomica* which was published in 1564.

the middle of the sixteenth century. He gave excellent descriptions of the ear, including the eardrum, the muscle which tenses it, called the tensor tympani, the ossicles, the auditory tube, and parts of the inner ear.

Probably the first book to be devoted solely to the ear was Eustachius's *The Examination of the Organ of Hearing*, which was published in 1562. Although the auditory tube was known to the ancients and had actually been dissected by Alcmaeon and described by Aristotle, Eustachius's classical description appeared in this book. Eustachius, who had dissected the inner ear, also elucidated earlier descriptions of the cochlea and of the tensor tympani muscle.

Giulio Casserio, better known as Casserius, and another of the great anatomy professors at Padua, specialized in the structure of the ear and the larynx. In 1601, he published an extensive account of the ear in both humans and animals. It was based on careful dissections and included illustrations of the ossicles and of the muscles and cartilage of the external ear.

One of the bright lights in the anatomy and physiology of the ear was Antonio Valsalva, who is best remembered eponymously for the sinuses in the aorta. His great work was on the ear, however, and his book *De Aure Humana*, published in 1704, was to remain a standard work on the subject until the nineteenth century. Valsalva, who in this book first called the auditory tube the Eustachian tube, was also the first to divide the ear into three sections—the external ear, the middle ear, and the inner ear. He carefully investigated the functions of the labyrinth and the tympanic cavity and described the sebaceous glands and lymph nodes, as well as the muscles of the external ear. Valsalva's ligament and muscle are both in the external ear and the great incisura of Valsalva is part of the ear flap, or pinna.

On the upper edge of the pinna, which was named by Rufus of Ephesus from the Latin for a wing, there is often a small projection which is an evolutionary remnant of the upstanding, or prick, ears of some animals. This projection, which is particularly prominent in some people, was considered by Charles Darwin, the great Victorian naturalist, to be of anthropological significance and it is known as Darwin's tubercle.

Thomas Buchanan, a British physician who practised in Hull, Yorkshire, in the nineteenth century, made a particular study of the outer ear and tympanic membrane and in 1828 published his *Physiological Illustrations of the Organ of Hearing*. One of the first to recommend the use of artificial light for examination of the eardrum, Buchanan devised a complicated lamp, using a candle as the means of illumination, which became known as the inspector auris.

A name in the middle ear well known to English-speaking anatomists is Shrapnell's membrane, which refers to the flaccid part of the tympanic membrane. Here, however, the man is more interesting than the structure, for Henry Jones Shrapnell was a native of the Gloucestershire village of Berkeley which became infamous for the site of the horrific murder of Edward II. Shrapnell, who was born in Berkeley towards the end of the eighteenth century, married Maria Marklove, ward of Edward Jenner, the country physician celebrated for his introduction of smallpox vaccination. Shrapnell, who was surgeon to the local regiment of militia in the county, described his membrane in 1832, two years before his death.

The same structure is sometimes called the membrane of Rivinus, recalling its description more than a century earlier by Augustus Rivinus, a botanist, physician, and professor of physiology at Leipzig. Here we have one of those complex family situations, for the membrane has been wrongly attributed to his son Johann Augustus Rivinus, who reported his father's observations in his inaugural dissertation in 1717. To complicate matters, there was also an Andreas Rivinus, the grandfather, and he, too, was professor of physiology at Leipzig. The Rivinus family is a prime example of the German medical dynasties which have caused havoc among bibliographers.

The Danish anatomist Ludwig Jacobson has a nerve, a canal, and a plexus of nerves, all of which run in association with the tympanic cavity, attributed to his name. Jacobson, who later served as military surgeon to the French army, wrote about these structures in the first years of the nineteenth century.

Also in the middle ear, the tensor tympani, which was familiar to sixteenth-century anatomists, is sometimes known as Toynbee's muscle after Joseph Toynbee, an English ear surgeon, who wrote about it in 1851. Toynbee and his

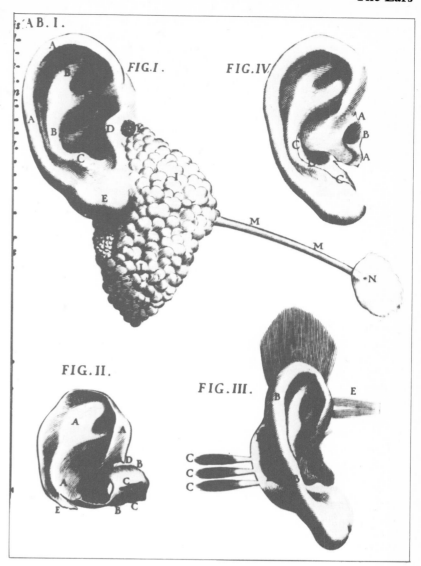

contemporary the Irish surgeon Sir William Wilde, the father of Oscar Wilde, were the pioneers of British otology.

Wilde, a distinguished surgeon and antiquarian and one of the leading figures in the medical life of Dublin in the mid-nineteenth century, made a special study of the anatomy of the tympanic cavity, recognizing this as the site of origin of many diseases of the ear. His name is linked with the cone of reflected light which is seen when artificial light is used to examine the eardrum. Wilde described this cone of light in 1853, at which time he would have had at his disposal an instrument devised by Toynbee in 1850, the forerunner of the modern otoscope, which was not developed until the 1870s. The shape of Wilde's cone, reflected back directly from the funnel-shaped eardrum, is diagnostic in some hearing disorders.

This same cone of light is known in Europe as Politzer's cone after Adam Politzer, the Hungarian-born otologist who, in his day, was regarded as the world leader in this speciality. Politzer, who was appointed the first professor of otology at Vienna in 1870, wrote about the cone nineteen years later.

The best known of the many eponyms in the inner ear is the organ of Corti, the spiral, sound-receiving apparatus in the cochlea and the vital part of the whole sound-detection system. Marchese Alfonso Corti was born in Sardinia in 1822. He was a histologist and developed various staining techniques. In 1852, he used carmine to show up the details of the contents of the cochlea. Although Corti also added to the microscopical knowledge of the retina of the eye, it is for the organ of the inner ear that posterity remembers his name.

Cladius cells, which are the supporting structures in the cochlear canal, recall a nineteenth-century professor of anatomy at Kiel and Marburg. Cladius cells also form part of Hensen's canal, which commemorates Viktor Hensen, best known for his work in the field of embryology, who was also at Kiel as professor of physiology in the late nineteenth century. Hensen's canal joins the cochlear canal to the saccule, a pouch-like structure in the inner ear.

There are many other eponyms in the cochlea, including Bottcher's ganglion, canal, and cells, Breschet's hiatus, the columns of Cotunnius, Gottstein's process, Loewenberg's canal, Nuel's space, Reissner's membrane, Rosenthal's canal, and Siebenmann's canals. All of these bear witness to the painstaking work of men who dissected the delicate structure of the inner ear.

It was not until the eighteenth century that investigators began to consider the inner ear as a resonant structure. One of the foremost of these was the Italian anatomist Domenico Cotugno. In 1761, he described the labyrinthine system and the fluid within it, and also propounded a theory of hearing.

The most systematic and brilliant work on the physiology of hearing was done in the nineteenth century by the great German physicist and physiologist Hermann von Helmholtz. After doing extraordinary work on the eye, he turned his attention to the ear. While he was professor of physiology at Heidelberg, between 1858 and 1871, he devoted himself to acoustics and, inspired by the anatomical studies of Corti and by Cotugno's concept of hearing postulated a century earlier, von Helmholtz elaborated a theory of hearing based on resonance. This became widely known as the piano theory, a name suggested by the analogy between the cochlear fibres and the strings of a piano.

Almost a hundred years later the resonance theory was developed further in the United States by the Hungarian-born physicist Georg von Békésy—work for which he was awarded the Nobel Prize in 1961. Today's knowledge of

In 1704, Antonio Valsalva published his great work on the ear entitled *De Aure Humana,* from which these illustrations are taken. Valsalva's book, in which the auditory tube was first referred to as the Eustachian tube, was to remain a standard work on the subject until the nineteenth century.

Discovering the Human Body

Charles Darwin overturned humanity's view of itself at the centre of the universe with his controversial theory of evolution. Born in Shrewsbury in 1809, he went to Cambridge to study for the ministry. The strongest influence he encountered there was that of the botanist John Henslow, who persuaded him to join *HMS Beagle* as expedition naturalist on her voyage in 1831 to South America and the Pacific. This was to transform Darwin into one of the finest biologists of the century. All his future work was based on the observations he made during this five-year passage. By 1837, Darwin had grasped the principle of natural selection. He believed that

species in nature were mutable and shared common ancestry, but it was another twenty years before he proved it. His work on evolution, *The Origin of Species,* was published in 1859—to a fiery reception from scientists, churchmen, and humanists, none of whom welcomed the notion that humans are descended from the apes. Darwin, a sick man in the years after the voyage—he is believed to have contracted a form of sleeping sickness in the Andes—modified his original theories in subsequent editions of *The Origin of Species.* Living almost as a recluse in later life, Charles Darwin died in 1882 and was buried in Westminster Abbey.

Marie Jean Pierre Flourens, the French anatomist, lived from 1794 to 1867. He was the first to prove that the cerebral hemispheres are the organs of sensation and will. Using pigeons as subjects, he removed portions of their brains and, in 1824, isolated the centres of nerve functions. He found that the cerebrum is the site of thought and will power, and that the coordination of muscle movements is located in the cerebellum. Four years later, he discovered the relation of the semicircular canals to vertigo. In 1837, he proved that the medulla is the centre of respiration. Flourens also did important work in toxicology and wrote on the history of the discovery of the circulation and on longevity.

how hearing works, however, is mainly due to the research of von Helmholtz.

Anatomists of the seventeenth and eighteenth centuries had no idea how the second and less obvious function of the ear, that of balance, was achieved. Although the positions of the three semicircular canals were well known, it was generally assumed that the labyrinth was wholly concerned with hearing and the canals were thought to be involved in sound perception.

The first person to show that the labyrinth is vital to equilibrium was the French experimental neurologist Marie Jean Pierre Flourens. In 1824, he described experiments in which he had produced abnormal head movements in a pigeon by cutting each of the semicircular canals in turn. The plane of the abnormal head movements was always the same as that of the severed canal. Flourens also demonstrated the distinction between the two branches of the auditory nerve. He showed that while hearing was not affected by destroying the vestibular fibres, it was lost completely when the cochlear fibres were cut. This indicated a second and separate function for the vestibular branch. Almost half a century was to pass before the significance of Flouren's work was appreciated and the semicircular canals were accepted as specific sensory units related to movements of the head.

In 1870, the German physiologist Friedrich Goltz developed his hydrostatic concept—that the canals are stimulated by the weight of the fluid, or endolymph, they contain, and that the pressure exerted by this fluid varies with the position of the head. Three years later, however, three scientists, each working independently—Ernst Mach and Josef Breuer, both Austrians, and Crum Brown, a Scots chemist—arrived at the hydrodynamic concept which is that movements of the head cause a disturbance of the fluid in the canals, which are receptive to the movement of fluid or to changes in pressure. Later research was to prove this hydrodynamic concept to be correct.

It is now known that normal equilibrium is the result of the labyrinth in each ear functioning reciprocally. When, however, the labyrinth in one ear is damaged, it is the continuing normal reaction of the other—which is unchecked by reciprocal controls—that causes such unpleasant effects as vertigo.

With its dual functions, one of which was not even discernible in the days when men believed that goats breathed through their ears, the ear has presented generations of physiologists with a challenge quite disproportionate to its compact size. And research is continuing into aspects of both functions—the physics of acoustics and the fine mechanisms of balance deep inside the inner ear.

A hand cupped behind the ear, directing more of the initial sound to the ear, is probably the earliest attempt to "aid" hearing. It is easy to imagine the step from the hand to a hollow animal horn or a broken shell. The first historical references to hearing aids are to this type, and to large leaves or lengths of cane which would concentrate sound energy through a limited pathway to the ear. For centuries, the material varied— horn, wood, metal, shell, but the principle stayed the same—simple amplification. Only the shape and ornamentation were subject to variation. Some strange objects found near Pompeii, shaped like funnels with a spiral tube at the smaller end, many of them foldable, were Roman hearing aids.

The effort to find more effective, and more discreet, aids resulted for a long time mostly in more and more baroque shapes. The problem of needing one hand to hold them was overcome by a kind of headphone arrangement which came in monaural and binaural models. There was considerable experimentation with bone conduction, in which the sound is applied to the skull rather than by way of the ear mechanisms.

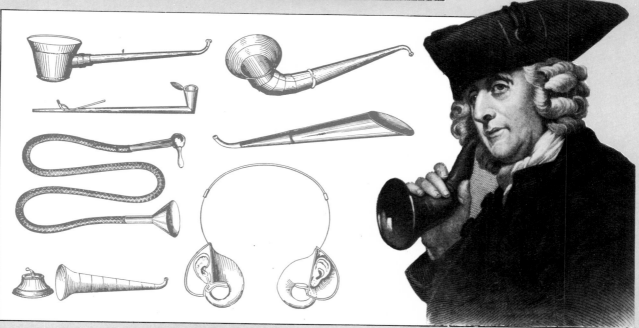

Bone-conduction devices were awkward. They usually had to be held in the teeth of the listener and those of the speaker. Sophisticated bone-conduction devices have now been developed. Modern hearing aids still only amplify sound, but they do so with power and clarity, and are differentiated to suit many kinds of hearing loss. They are electrical or even electronic, and with the advent of the micro chip will become even smaller and more efficient.

The Eyes

In this illustration from *De Fabrica,* Vesalius perpetuated the Galenic misconception of the lens being at the centre of the eyeball.

For many centuries ignorance about the nature of light made it impossible for the functioning of the eye to be understood. In ancient times, the main stumbling block was the concept that vision was the result of something coming out of the eye to impinge on the object looked at, instead of exactly the reverse.

Alcmaeon, the physician and anatomist from the Greek colony of Croton in southern Italy, was one of the first to mention the eye. In about 500 B.C., he described the optic nerves and

decided that three things were necessary for vision—external light, the "fire" in the eye, and the liquid in the eyeball. It is not clear from these early fragmentary writings whether his reference to external light meant that he had an inkling that light entered the eye, rather than that a beam was projected from it. Certainly the projection concept lasted much later than the theories of the Greeks, who had a fair knowledge of optics generally. Alcmaeon assumed that there was "fire" in the eye because he knew that

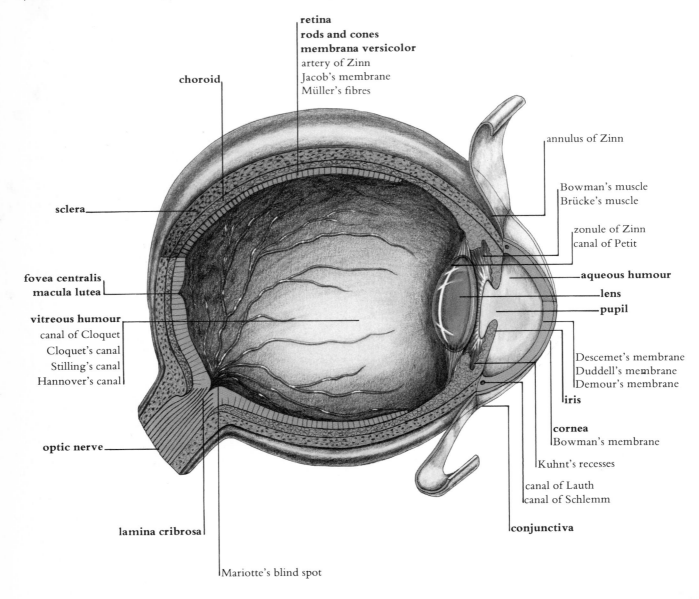

retina
rods and cones
membrana versicolor
artery of Zinn
Jacob's membrane
Müller's fibres

choroid

annulus of Zinn

sclera

Bowman's muscle
Brücke's muscle

zonule of Zinn
canal of Petit

fovea centralis
macula lutea

aqueous humour

lens

pupil

vitreous humour
canal of Cloquet
Cloquet's canal
Stilling's canal
Hannover's canal

Descemet's membrane
Duddell's membrane
Demour's membrane

iris

optic nerve

cornea
Bowman's membrane

Kuhnt's recesses

canal of Lauth
canal of Schlemm

lamina cribrosa

conjunctiva

Mariotte's blind spot

a blow on the eye produces sparks—the experience of seeing stars. The liquid he referred to is the jelly-like vitreous humour that he found when he dissected animals and recognized as necessary for the transmission of sight.

Hippocrates, who lived about a century after Alcmaeon, was more concerned with diseases of the eye, but in the Hippocratic writings it was theorized that sight occurred from the formation of the image on the pupil, rather than on the retina, as actually happens. Operations for the removal of cataracts were among the surgical techniques performed in those early times, but in the theory of vision there was no mention of the soft, transparent lens which adapts itself in shape to permit the eye to focus.

Rufus of Ephesus, who was a student at Alexandria in about A.D. 50, was responsible for introducing most of the names still used for various parts of the eye. Although the great medical school was in decline by this time, compared to its heyday three centuries earlier when Herophilus and Erasistratus worked there, Rufus produced a number of books which disappeared via the Arabic route after the fall of Rome and were unknown in Europe until 1554. Naturally, Rufus wrote in Greek, but his descriptions are used as straight translations either from the Greek or the equivalent given word in Latin.

Rufus's description of the eye was sufficiently detailed to form the basis of an anatomical

diagram, the main features of which are remarkably accurate. The parts he named include the cornea, which means horny, an apt description of this tough, transparent membrane which is like a window in front of the eye, and the lens, named from the Latin for a lentil, because of its shape. Rufus used the word conjunctiva, from the Latin to conjoin, to describe the fine membrane which covers the front of the eyeball and serves as a lining between the eyeball and the inner lids. His knowledge of the conjunctiva was inexact, however, for he thought it extended all around the eyeball. He applied the term sclerotic, meaning hard, to the tough, fibrous capsule of the eyeball, what is now known as the sclera, or white, of the eye. The next membranous layer, which is vascular and rather spongy, he called the choroid—from chorion, the name of the lining of the pregnant womb.

Perhaps because of its irridescence, the muscular, pigmented disc which gives the eye its colour suggested the name iris to Rufus, from the word meaning rainbow in Greek, and, since a rainbow was regarded as a sign or announcement from the gods, it was personified in Greek mythology as Iris, the messenger from the gods to mortals.

The multilayered retina, which lines the eyeball and carries the actual light-sensitive elements, the rods and the cones, gets its name from a net, and the clear jelly of the major part of the eyeball appeared vitreous, or glass-like, to Rufus. He also named the optic nerve, which runs from the back of the eyeball to the visual cortex in the brain, from the Greek verb to see. Thus the nomenclature of the eye laid down by Rufus has survived virtually intact almost since the beginning of the Christian era.

Operations for the removal of cataract were among the earliest of surgical procedures, with techniques dating back to Hippocratic times. This illustration, from an English manuscript of the early thirteenth century, shows a surgeon removing a cataract.

Islamic in origin, this schematic view of the eyes and brain, which shows the optic nerves crossing at the optic chiasma, is the oldest surviving drawing of its kind, dating back perhaps to A.D. 1000. Its geometric form was probably dictated by Moslem resistance to lifelike portrayals of human anatomy.

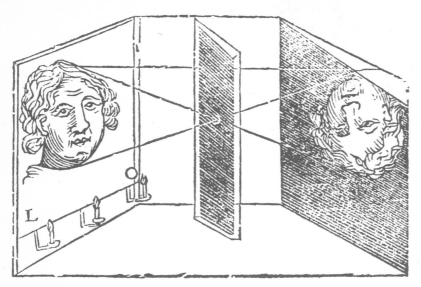

This seventeenth-century illustration of the principle of the camera obscura shows how the image of the lighted face on the left appears inverted on the wall of the darkened chamber on the right. It was the early camera obscura that suggested the role of the optic nerve in vision to Leonardo da Vinci.

Soon after Rufus's time, Claudius Galen emerged as the dominant figure in the Roman medical world. He dissected the muscles of an ox's eyeball and stated that the optic nerves were the first of seven pairs of cranial nerves. He was entirely wrong, for there are twelve not seven pairs of nerves and the optic nerves are the second pair, since the nerves of smell lie in front of them. Galen made little contribution to knowledge of the internal structure of the eye, but caused some confusion by insisting that the lens was exactly in the middle of the eyeball, rather than in front of the eye and assumed, as others had before him, that the lens itself was the recipient of vision.

Galen's error, along with much of the ancient anatomical descriptive material which returned to Europe circuitously through Moorish translations, reappeared centuries later in the writings of such anatomists as Mondino, who worked at Bologna in the early fourteenth century. As did all the Arabs who followed Galen's original mistake, Mondino placed the lens of the eye in the centre of the orbit. This error was perpetuated for centuries.

Even Leonardo da Vinci, who knew more about vision than anyone up to his time, drew the lens in the wrong place. As the result of his knowledge of optics, however, he decided that the optic nerve at the back of the eye, rather than the lens, was the place where the reception of light took place, because of the inversion of the image. By then, the camera obscura was known—Leonardo actually experimented with it—and it was realized that light came in through the pinhole and gave an inverted picture at the back of the box. Leonardo also noticed that, particularly in nocturnal animals, the iris constricted on exposure to light. He believed that the sharpness with which an object was seen depended on the intensity of illumination and the size of the pupil.

Vesalius was less than brilliant when it came to the eye. He described the gross anatomy in an indifferent fashion and included in *De Fabrica* a poor diagram of the internal structure. The lens was strung halfway across the eyeball, and the anterior chamber, which contains the watery fluid known as the aqueous humour, was shown as equal in size to the vitreous humour, the transparent gel which occupies the back of the eye. It is difficult to understand how he made this error since the most cursory examination of an eyeball, animal or human, would have revealed the true structure.

Slavishly following the long-dead Galen, Vesalius declared that the lens received the optical image. Yet, since it must have been obvious to Vesalius that the lens has no connection with the optic nerve, whose only function could be to convey impressions to the brain, it is again a tribute to the power of Galen that his theory overrode the observational powers and keen intelligence of such a man.

Realdus Columbus, the pupil and successor of Vesalius in the chair of anatomy at Padua, was one of the first to refute the Galenic belief that the lens was in the centre of the eyeball. He placed it in its correct position behind the iris. Columbus also accused his former master of dissecting the eye of an ox, rather than that of a human. It is strange today, when there is some semblance of professional etiquette, to learn how men of science in past centuries would publicly revile and abuse their rival colleagues, whose feelings seemed to be less easily bruised than those of their modern counterparts.

The Swiss anatomist Felix Plater, who lived between 1536 and 1614, also relocated the lens correctly. In his book, which was published in Basel exactly forty years after the publication of Vesalius's *De Fabrica,* Plater also ascribed the correct function to the retina. A professor of medicine at Basel for forty years, Plater had dissected more than three hundred bodies and was the first disciple of Vesalius north of the Alps. He founded a botanical garden and an anatomy theatre modelled on the one at Padua.

Fabricius of Aquapendente, one of the greatest teachers of anatomy at Padua and a contemporary of Plater, wrote a book on the eye and vision in which there were the best illustrations up to that time of the interior structure of the eyeball. Fabricius was the first to understand the true construction of the lens, but, unlike Plater, who wrote at an earlier date, he failed to place the seat of vision in the retina, thinking along the same old erroneous lines that the lens itself received the image.

René Descartes, the philosopher-scientist who expounded the so-called Cartesian theory that the body was merely a complex mechanical device, also wrote on the theory of sight. In his book *De Homine,* published in 1662, he gave

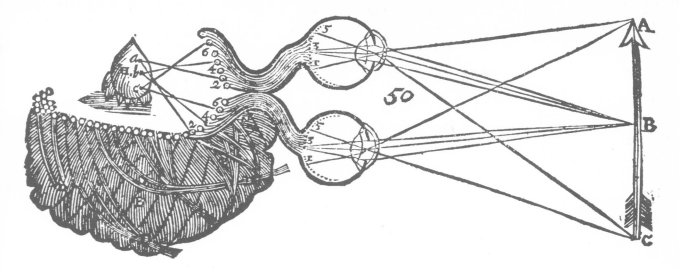

explanations and included drawings of the visual system, one of which showed an inverted image focussed by the lens on to the retina. From then on, his credibility began to waver for he postulated that the optic nerves were hollow tubes connected to the walls of the brain ventricles and that animal spirits passed back to the pineal gland and hence to the muscles. Descartes was partly correct about the initial stages of the visual process and about the reflex arc, but the nub of his argument was still pure Galen.

In 1749, Albrecht von Haller, the distinguished Swiss anatomist, botanist, and physiologist, described the choroid layer of the eyeball and the lamina cribrosa, the "perforated plate" on the rear wall of the eye where the optic nerve fibres make their exit.

The retina, the vital part of the eye, which picks up the light and converts it into electrical nerve impulses, much as a television camera functions, has a complex structure. In 1668, an interesting discovery about the retina was made by Edmé Mariotte, Prior of the Cloister of Saint Martin sur Beaune near Dijon in France. A physicist as well as a cleric, and one of the founder members of the French Academy of Sciences, Mariotte wrote on all manner of scientific subjects, including the motion of fluids, the nature of colour, and the freezing of water. In 1668, this polymath published a paper on what became known as Mariotte's blind spot. The blind spot is the tiny area on the retina where the optic nerve leaves. Because there are no vision receptors in the form of rods or cones here, there is a defect in the visual field. This can easily be demonstrated by drawing two dots about four inches (10 cm) apart. If one eye is closed and the other eye fixed on one spot, the other spot will disappear when the paper is slowly moved away from the face.

In the eighteenth century, the microscopic structure of the lens was described by Johann Reil, the German physician who is best known for the island of Reil in the brain. Reil also fully described the macula lutea, the focal point of fine vision which is the antithesis of Mariotte's blind spot.

In the macula lutea, the yellow spot, visual acuity is greatest owing to a dense crowding together of the colour-sensitive cones, especially at the centre, which is called the fovea centralis, the central pit. The overlying layer is almost absent, therefore a distinct image can be kept in focus on the almost bare receptor cells. It is this area that is normally used to view the world. The eyes are continually moved to keep the object being studied in focus on the macula. The rest of the retina receives a less well-defined picture.

Located away from the macula, the rod receptors, which are concerned with black and white vision, have better receptivity in dusk or gloom than the colour-sensitive cones of the fovea. Consequently, peripheral vision is more efficient in poor light.

Jacob's membrane, the important layer in the retina containing the light-sensitive rods and cones, was described in 1819 by Arthur Jacob, a Dublin eye surgeon and professor of anatomy and physiology.

In 1832, George Fielding, an English ophthalmologist, described a new membrane in the eye, a retinal layer now known as the membrana versicolor.

The supporting nerve fibres, or neuroglia, of the retina became known as Müller's fibres following their description in 1856 by Heinrich Müller, the German anatomist who is also commemorated in Müller's muscle, which lies around the outside of the eye socket.

In 1775, the Abbé Felix Fontana, the distinguished anatomist, physiologist, and naturalist, who founded the Florentine Museum of Natural History, described the movements of the

In his book *Principia Philosophiae*, published in 1644, René Descartes illustrated his eccentric theory of vision. Although he was correct in showing an inverted image which was focussed by the lens on to the retina, he then went on to show the optic nerves as hollow tubes linked to the brain ventricles, from where animal spirits passed to the tiny pineal gland in the brain and from there to the muscles.

Thomas Young, the English doctor who only lived for thirty-six years, displayed his genius at an early age. Born in 1773, he learned to read when he was two years old. While he was still a medical student his interest in sensory perception led to his discovery of visual accommodation. When he was twenty-one he was elected to the Royal Society for his work in optics. Young's interest in physics led him to reject Newton's theory of light as particles and to establish the wave theory of light. An authority on the language of Egypt, he helped to decipher the hieroglyphics on the Rosetta stone.

ciliary muscle which alters the shape of the lens. Young published a book on ophthalmology in 1801 in which he described various optical defects and also put forward a theory of colour vision; he suggested that the retina is sensitive to red, green, and violet. (Earlier, in 1794, John Dalton, the chemist, had written on colour blindness from which he suffered and which has been known as Daltonism ever since.)

One of the most important accounts of the physiology of sight was the *Handbook of Physiological Optics*, written by Hermann von Helmholtz, and published over ten years, from 1856 to 1866. In this von Helmholtz supported Young's theory of colour vision, which is now known as the Young-Helmholtz theory. This remarkable German scientist, equally distinguished in physiology, mathematics, and physics, invented the ophthalmoscope in 1851 and was the first person to observe the living human retina.

muscles of the iris in controlling the entry of light to the eye. The iris has both circular and radial fibres which regulate the size of the pupillary opening and, therefore, the amount of light to fall on the retina.

In the middle of the eighteenth century Albrecht von Kölliker, the Swiss anatomist, described the iris and theorized correctly that it functions by smooth, involuntary muscle, which responds reflexively to the intensity of light.

In 1792, Thomas Young, a pupil of the famous anatomist and surgeon John Hunter, published his theory that accommodation (the ability of the eye to focus on objects at varying distances) is accomplished by alteration of the radius, or curvature, of the flexible lens. This is correct, although Young was mistaken in thinking that the lens itself has muscular powers, when actually it is the contraction of the surrounding

The sensory organs seem to have attracted a far greater proportion of eponyms than more general anatomy, and in this respect the eye is no exception. There were at least sixty named parts of the eye and its associated muscles and lacrimal apparatus on record before the revision of nomenclature swept many of them away. Some of the more important ones include Descemet's membrane, which is the delicate layer of cells that lines the back of the cornea, the window in front of the eye. This was described by Jean Descemet, professor of anatomy and surgery in Paris, in 1758, when he was only twenty-six years old. This same membrane had been mentioned earlier, in 1729, by an obscure English surgeon named Benedict Duddell. To complicate the matter still further, Pierre Demours, a Marseilles ophthalmic surgeon with many discoveries to his credit, also published a paper on this membrane in 1767. It is, therefore, also known as Duddell's membrane or Demour's membrane, although Descemet is most frequently commemorated.

Sir William Bowman, who, in spite of his other anatomical and physiological interests was one of the great pioneers of the eye, has several attributions there, in addition to his more famous one in the kidney. Bowman's membrane is the elastic sheet in the middle layer of the cornea. Bowman's muscle is the ciliary muscle, which alters the curvature of the lens in accommodation, or focussing. It was described by Bowman in 1847, but is sometimes called Brücke's muscle, after Ernst Brücke, a professor in Königsberg and then in Vienna, who independently published on the ciliary muscle in the same year.

In the unborn child, the retinal artery, which comes up the centre of the optic nerve, continues across the centre of the eyeball to supply developing structures at the front of the orbit.

John Dalton, the self-taught English physicist and chemist, was born in 1766, the son of a Quaker weaver. In 1793, he began to teach mathematics and physics at the New College of Manchester. Dalton did his early research on the aurora borealis—the colours of which he was unable to see—and on tradewinds. Dalton had remarkable gifts of intuition and observation. He developed a system of chemical notation, applied the atomic theory to chemistry, and developed the primary laws of heat and gases, including Dalton's law. From 1817 until his death in 1844, he was president of the Manchester Philosophical Society.

A fifteenth-century painting recalls the legend of St. Lucy, patron saint of vision, whose martyrdom is symbolized by a pair of eyes blossoming on a twig. St. Lucy, who lived in Syracuse during the reign of Diocletian, was betrothed but chose to forego marriage. On learning that her eyes continued to haunt her disappointed suitor, she is said to have torn them out and sent them to him on a platter.

This part of the vessel later shrivels, so that it does not impede the passage of light to the retina. A ghost of it is usually left behind and this can sometimes be seen as a "floater" when staring for any length of time at the sky or at a blank wall. Extending across the centre of the vitreous jelly is a narrow canal for the defunct artery. This was named for Jules Cloquet, one of two brothers who were professors of anatomy and surgery in Paris in the nineteenth century. Jules, the younger brother, described the canal in 1818. (His elder brother, Hippolyte, had a nerve ganglion named after him.) A weak rival eponym for Cloquet's canal is Stilling's canal, named for Bernard Stilling of Vienna who wrote about it fourteen years after Cloquet.

Johann Gottfried Zinn, professor of medicine at Göttingen in Germany, published a classical treatise on the eye in 1755 and, consequently, is commemorated three times. The artery of Zinn is the central artery of the retina. The annulus of Zinn is the ring at the back of the eyeball to which the muscles that move it are attached. The zonule of Zinn is a tiny membrane at the edge of the lens.

The tiny channel between the zonule of Zinn and the vitreous jelly has two names—the canal of Petit and the canal of Cloquet. But here Jules Cloquet loses any claim for François Petit published his description in 1726 and in 1755 Zinn referred to it as Petit's canal. Petit, who was born in 1664, worked mainly in Paris and was reputed to be an exceptionally skilful ophthalmic surgeon. Yet another claimant for this same little canal was Adolph Hannover, a Danish anatomist who had studied in Germany under Johannes Müller. Hannover's claim is tenuous, since he did not describe the canal until 1845.

Leading off this popular canal are little cavities known as Kuhnt's recesses. They are named for Hermann Kuhnt, a German anatomist and ophthalmic surgeon who practised in the late nineteenth and early twentieth centuries.

Where the transparent cornea meets the opaque sclerotic coat of the eyeball another well-known canal exists. This is almost universally known as the canal of Schlemm, although a slightly earlier description by a professor in Strasbourg entitles it to the alternative name of the canal of Lauth. Ernest Lauth described the canal in 1829, while Friedrich Schlemm, professor of anatomy at Berlin, published his description a year later, yet captured the eponym almost completely.

Outside the eye, the lacrimal apparatus which produces and disposes of the tears has many eponyms, but some of these are in the nose. The function of the tear gland, situated beneath the outer end of the upper eyelid, is to wash the delicate cornea with a clear, sterile fluid, to keep it clean and free of infection. The gland is like the washers of a car window and the lids are like the wiper blades, blinking involuntarily every minute or two to sweep the corneal window clear and bathe it in lacrimal fluid. The excess is drained off through two small apertures on the lids at the inner end, passing down through the tear ducts to the nasal cavity. Many animals have a special third eyelid for this

In the middle of the nineteenth century, as shown in this illustration which appeared in a French journal in 1885, the newly invented, if still crude ophthalmoscope permitted physicians to examine the interior of the eye.

purpose, called the nictitating membrane, of which the pink fleshy substance in the corner of the human eye is the only vestige.

The eyelids are protective shields which are lowered involuntarily by means of a powerful reflex action when the eye is touched or even threatened. Sleep also induces the lids to drop, to prevent drying during a long period of inactivity. Some fibres of the circular muscle which runs around the whole eye socket—the orbicularis oculi muscle used to screw up the eyes—pass across the lacrimal sac at the inner end of the eye. The sac is a collecting reservoir for spent tears. This muscle is known as Horner's muscle after William Horner, professor of anatomy at the University of Pennsylvania, but his description in 1824 was long anticipated by Joseph Duverny, the founder of the Paris school of anatomy, who wrote on the eye muscle in 1749. Horner is better known, however, for Horner's syndrome, an affliction in which the eyelid droops because of weakness of the muscles.

Another noteworthy eponym in the eyelids is the Meibomian glands, which lie in a row along the edge of the eyelid. These glands are significant because blockage and infection may cause Meibomian cysts, which can be painful and persistent. Heinrich Meibom, better known by the Latinized name Meibomius, was born in northern Germany in 1638, the son of an anatomist. A prime example of the polymath, he became professor of medicine, history, and poetry in Helmstadt, and wrote about the glands of the eyelid in 1666. Giulio Casserius, however, one of the long line of anatomists at Padua, wrote the first description of these glands almost sixty years earlier.

The ciliary glands, which lie between the roots of the eyelashes, have been known at various times as Moll's, Sattler's, or Zeis's glands. Jacob Moll was a Dutch eye doctor from the Hague and he described the ciliary gland twenty years before Hubert Sattler of Leipzig wrote about them in 1877. Edouard Zeis of Dresden and Marburg anticipated both with his description of the glands in 1835.

Although Rufus of Ephesus described the anatomy of the eye in accurate detail in about A.D. 50, it was not until the late eighteenth century—when the physics of optics and technical knowledge of optical instruments was well established—that the visual process itself began to be understood. From the middle of the nineteenth century many discoveries were made as the result of the efforts of ophthalmic surgeons to correct visual defects. And the minute histological structure of the eye is still being investigated today by such modern means as electron microscopy.

The physician Rhazes, left, the best-known and most prolific Islamic medical writer, lived from A.D. 865 to 925. Ophthalmology was one of the fields in which Arab physicians were most active during their long custody of classical medical learning. No less than thirty Arab texts on the eye survive from this period. Rhazes is said to have lost his vision after being beaten on the orders of the ruler of Bokhara. Later, seeking to regain his sight, he had the chance of an operation, but declined when he found the surgeon knew nothing of the anatomy of the eye.

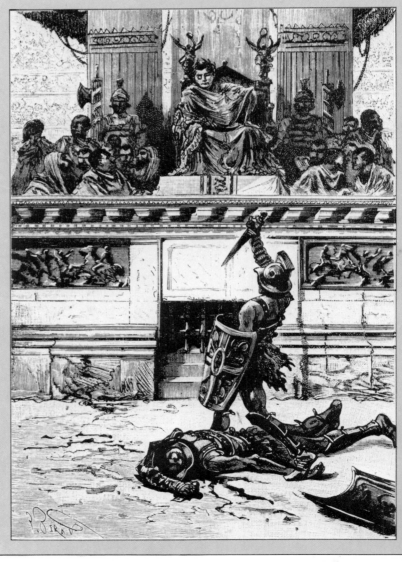

The use of simple lenses as magnifiers has been known in the West for more than a thousand years. Nero wore a concave emerald in one eye as a makeshift eyeglass to help him follow the gladiatorial combats. In the early thirteenth century, Bacon said of the magnifying lens, "This instrument is useful to the aged and to those who have weak eyes." Marco Polo returned from China with reports on the use of lenses there which sparked off much interest. A medical text of 1300 mentions spectacles.

The first painting in which spectacles are represented is the portrait reproduced right, of Cardinal Hugues painted by Tommaso di Modena in 1360. The spectacles are primitive pince-nez, no doubt with simple magnifying lenses. A possible first appearance of the word in literature is Chaucer's use of "spectacle," in the *Wife's Tale*, as something to see through.

For reasons which remain obscure, since most of the population was illiterate, by the mid-sixteenth century spectacles, whether they were necessary or not, enjoyed an enormous vogue, and seem to have been available on every street corner. The popular craze for eyeglasses is wryly captured by the engraving below, the work of the sixteenth-century artist Johannes Stradanus. Clearly everyone but the children is caught up, but only the two men in the right-hand corner are obviously using their glasses.

Spectacles were made and sold in this haphazard fashion for centuries, and it was not until optical processes and structures were clearly understood that eyeglasses could be made to begin to adequately correct a wide range of visual defects.

Blood and Lymph

Before the development of the microscope, blood was considered an indivisible fluid, mostly spilled in battle. This sixteenth-century chart *The Wound Man*, was used as a guide to the location and nature of battle injuries.

Two of the body's tissues are liquid—the blood and the lymph. The blood is really the parent of the lymph. A clear fluid which drains from the spaces between the body cells, lymph, like the other fluids in the body, including the cerebro-spinal fluid, the endolymph of the internal ear, and the synovial fluid of the joints, is derived from the liquid supplied to the tissues by the blood. The essential difference between blood and lymph is the presence in blood of red corpuscles, or erythrocytes, and plasma proteins.

All the blood in the body is—or should be—confined within its blood vessels, but the smaller molecules, such as water, can seep through the capillary walls and thus maintain tissue pressure and nutrition and the many other functions that the moist internal environment of the body demands. The excess fluid returns to the system by draining into the lymphatics, which eventually combine to rejoin the blood vessels by way of the thoracic duct in the chest.

None of this was known to the ancient men of

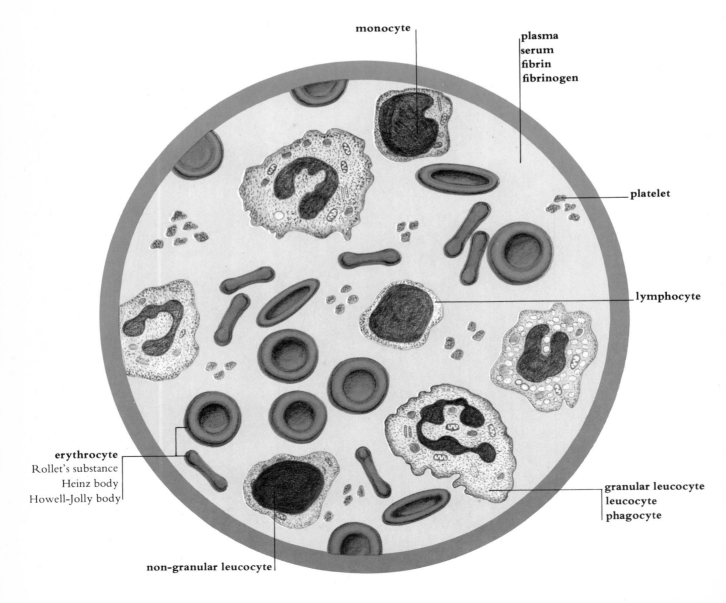

monocyte

plasma
serum
fibrin
fibrinogen

platelet

lymphocyte

erythrocyte
Rollet's substance
Heinz body
Howell-Jolly body

granular leucocyte
leucocyte
phagocyte

non-granular leucocyte

medicine, although many of the structures which could be seen with the naked eye were apparent to them. The lymphatics, or lacteals, of the intestine, for example, which become prominent after a meal because of the fat absorbed from food, had been described by the Greeks and Alexandrians, particularly by Herophilus and Erasistratus. But even after the time of William Harvey's discovery of the circulation, the exact function of the blood itself was obscure.

Although Harvey himself had stated that "the blood is the first to live and the last to die," it was not until the development of the microscope that the components and functions of the blood as a tissue—rather than as a hydraulic fluid—could be understood.

Today it is known that the adult human body contains ten to twelve pints of blood, and that suspended in the plasma, the part that is fluid, are the major components of blood—the red cells or corpuscles, known as erythrocytes, various types of white cells, or leucocytes, and the small particles known as platelets.

Red corpuscles are most numerous. Each cubic millimetre of blood—the size of a large grain of sand—contains five million red corpuscles, making a total of about twenty-five million million red corpuscles in the adult body. Three million new red cells are manufactured by the bone marrow every second, for the life span of an erythrocyte is only about one hundred days. The red cell is then destroyed in the spleen and its haemoglobin, the oxygen-carrying pigment which gives the red cells their colouration, is recycled to save the iron it contains. The paramount role of the red cells is to transport oxygen throughout the body.

There are about seven thousand white cells, of a number of different types, per cubic millimetre of blood. White cells are made in bone marrow and also in the spleen and lymph nodes. Many of

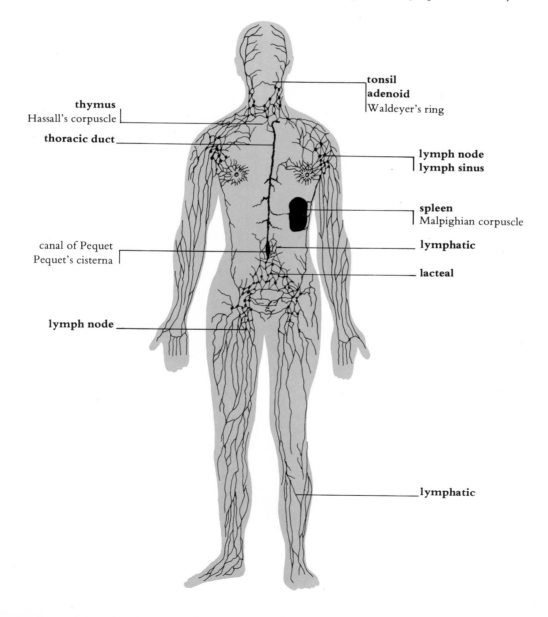

thymus
Hassall's corpuscle

thoracic duct

canal of Pequet
Pequet's cisterna

lymph node

tonsil
adenoid
Waldeyer's ring

lymph node
lymph sinus

spleen
Malpighian corpuscle

lymphatic

lacteal

lymphatic

Jan Swammerdam was born in Amsterdam in 1637. He studied medicine at Leyden with Stensen and de Graaf. Although he qualified, he never practised medicine, but devoted his time to research instead. Swammerdam became an expert microscopist and made extensive observations. He observed the red blood cells in 1658, discovered the valves of the lymphatics in 1664, and three years later described atelectasis, the incomplete expansion of the lungs at birth. At the end of his life, he suffered from depression. Shortly before his death in 1680 he burned his manuscripts in a fit of madness.

Anton van Leeuwenhoek, the Dutch linen draper, was an amateur scientist with no particular training. Born in 1632, he was established in his trade at the age of twenty. As a draper he had to use a simple magnifying glass to count threads. He learned how to use a simple, biconvex lens to magnify small objects. He became adept at grinding lenses with which he could see an astonishing number of things. His patience and attention to detail were rewarded by his observations of bacteria, his "little beasties." His hobby brought him fame— and a visit from the Queen of England and from Czar Peter the Great. In 1680, he was elected a fellow of the Royal Society. Leeuwenhoek ground about four hundred lenses in his lifetime.

Although he left much of his collection to the Royal Society, he kept his method of illumination a secret. He died at ninety.

physician, brought out a Latin edition in 1738, fifty-eight years after Swammerdam's death.

By that time, other men had discovered the red cell. In 1665, Marcello Malpighi noted these erythrocytes, but he did not realize their significance. The real credit for their discovery is always given to the archpriest of microscopy Anton van Leeuwenhoek of Delft, who gave an accurate account of them in 1674. He even made a close estimate of their size—equivalent to seven-and-a-half microns, a micron being one-thousandth of a millimetre. Leeuwenhoek also tried to measure the velocity of the flow of red cells through the small blood vessels. He studied blood cells from humans and animals and found that most mammals have circular red cells, while those of frogs, fish, and birds are oval. He correctly attributed the red colour of blood to the multiplicity of what he saw as pale yellow corpuscles, for he had to examine the erythrocytes without benefit of the stains used today.

It is difficult to imagine how this amateur microscopist, a linen merchant in a small Dutch town, was able to see such detail through the single lens of his crude, home-made instruments. He ground his own lenses, once even grinding a tiny one from a single grain of sand. Over a period of fifty years, Leeuwenhoek, the draper who wrote only Dutch and therefore could not publish in the Latin used almost exclusively by academics of the time, made all his scientific communications in the form of letters to the Royal Society in London.

Like most early observers, Leeuwenhoek thought that the red cells were spherical, whereas they are actually biconcave discs—like doughnuts with filled-in centres. It was Jean Senac, a French physician, who discovered the true shape of the erythrocytes in 1749, although the fact of the flat red cell was not generally accepted until the middle of the nineteenth century.

Twenty-six years after he had described the red cells in human blood, Leeuwenhoek noted a clear structure in the centre of the blood corpuscle of a fish. This was the nucleus of the cell. Unfortunately, for long afterwards others assumed they could see the same thing in human red cells. The possibility that no such nucleus existed in human erythrocytes was not even considered until 1827, when Joseph Lister and Thomas Hodgkin wrote a paper about cells. Actually, the erythrocyte is the body's only living cell which does not have a nucleus. It survives for about three months in this state, acting as a passive oxygen carrier.

Because of the poor definition afforded by early microscopes, until Lister and Hodgkin wrote about cells the body tissues were conventionally described as a collection of "globules." Swammerdam had simply called the red cells *particulae sanguinis*—particles of blood. Leeuwenhoek had described them as "small

these live for only a few hours. The white cells, which are concerned in the body's ceaseless fight against infection, scavenge bacteria and other debris in the blood and also produce such immune substances as antibodies.

The first man to see a red blood cell was Jan Swammerdam, in 1658, when he was only twenty-one years old. He was an eccentric—a wanderer, a religious personality, and afflicted with melancholia. He got no credit during his lifetime for the discovery of red blood cells. His descriptions and superb illustrations of all manner of microscopic life were published in Dutch in his *Bible of Nature*, which remained obscure until Hermann Boerhaave, the great Dutch

globules driven through a crystalline humidity." Lister invented better lenses and, consequently, he and Hodgkin were able to see the great variability of cell shapes, including the cells in the blood.

Within the red cells, several tiny structures have been named for their discoverers. Most of these are the remnants of the nucleus which disintegrates before the cell is released from the bone marrow into the circulation. When production is rapid, however, fragments of the nucleus may persist.

The best known of the nuclear remnants are the Howell-Jolly bodies, named after two quite independent observers. The first was an American, William Henry Howell, a professor of physiology at Johns Hopkins Medical School and later director of its School of Hygiene. Howell, who told the life history of all red corpuscles in 1890, described the bodies in 1905. Eighteen years later, the bodies were mentioned by Justin Jules Jolly, director of the histological laboratory and professor of physiology at the College de France in Paris.

Heinz bodies, small grains visible only with certain stains, are also presumed to be nuclear remnants. They were first described in 1904 by Robert Heinz, a professor of pharmacology at Erlangen in Germany.

Sometimes, actual nucleated red cells, known as normoblasts, appear in small numbers in the peripheral blood. Their appearance usually follows some phase of brisk production caused by haemorrhage or massive destruction of red cells. These interlopers are sometimes called Neumann's cells, commemorating Ernst Francis Neumann who drew attention to them in 1863. A professor of pathological anatomy at Königsberg, he also found where the red cells are manufactured in the depths of the bone marrow.

The featureless background material of the red cell were once known as Rollet's substance after Alexander Rollet, a nineteenth-century Austrian professor from Graz.

The red cells themselves were sometimes named after Giulio Bizzozero, a professor of pathology in Turin in the nineteenth century and the teacher of Camillo Golgi, the Nobel-winning neurohistologist. Bizzozero was Italy's premier haematologist.

A number of blood cells are named after Moritz Loewit, a professor of pathology at Innsbruck. The best known is the erythroblast, the parent of the red cell which lives in the bone marrow. Loewit described it in 1894.

The red blood cells are destroyed in the spleen, but some fragment in the blood vessels and the pieces are sometimes known as Arnold's bodies, after Julius Arnold. The son of the better-known Friedrich Arnold, who has many neurological structures named after him, Julius, like his father, was a professor of medicine at Heidelberg.

It was the French physician Jean Senac who, in the middle of the eighteenth century, first mentioned the white blood cells, describing them as "white globules of pus" which he had observed in the lymph. But it was not until 1855 that the word leucocyte for the basic white cell was first introduced in Littré and Robin's French dictionary of medicine.

In the eighteenth century, the pioneer haematologist William Hewson worked out the structure of the lymphatics in animals and investigated the role of the leucocytes. Hewson, a pupil and assistant of John and William Hunter, the famous anatomists and surgeons, wrote a book entitled *Experimental Enquiry into the Properties of the Blood*, which was published in 1771.

Hewson described the white cells in some detail, declaring that they are formed in the lymph glands and thymus and reach the bloodstream by way of the thoracic duct, the final lymph channel. He erroneously thought, however, as did many others, that red cells were changed to white cells in the spleen.

In 1843, Thomas Addison, one of the greatest British clinicians, wrote on the white cells. He described the escape of these mobile cells from the blood vessels into the tissues when they are attracted by infection or inflammation. He called this diapedesis, meaning leaping through, which more or less describes the way the leucocytes squeeze through imperceptible gaps between the cells of the vessel wall then ooze into the infected area, where millions of them attack the bacteria and form the yellow substance which is pus.

Soon after this, Wharton Jones, another British investigator, divided the white cells into two main types, granular and non-granular, a distinction which is still in use today. In 1846, Jones also described the amoeboid movements of the granular cells, which act as scavengers and engulf bacteria. The mode of action of these phagocytes, which means cells that devour, was elaborated in 1884 by Elie Metchnikoff, the Russian immunologist. He devised the theory of phagocytosis while working with sea creatures at a marine biology station in Sicily.

In 1883, Paul Ehrlich, who was to share the Nobel prize for physiology and medicine with Metchnikoff fifteen years later, traced the derivation of the granular leucocytes. He showed that these, like the red blood cells, are produced in the bone marrow.

All the blood cells and cell fragments float in the plasma, the featureless yellow fluid which is also the medium for the substances vital to clotting. Blood-clotting during life is called thrombosis, deriving from the Greek to curdle. The thin fluid which escapes from the clot is, logically, known as serum, from the Latin for whey, the watery residue left after milk has curdled. An embolism is said to occur when a thrombus, or clot, starts to move in the bloodstream and plugs a distant artery or vein, often with disasterous results.

Coagulation, a process not unlike the self-sealing facility in modern puncture-proof tyres, is, however, a vital defence mechanism in the body. Little defects occur all the time in the walls of the blood vessels and there is a continuous process of repair. When injury causes a major rupture, the platelets, the third element of the blood, adhere to the torn edges of the vessel, rapidly piling up to help bridge the gap. At the same time they release the activator substance which causes the plasma to clot, so that stratified layers, consisting of platelets, fibrin, and enmeshed red and white corpuscles, are laid down. Once the tear is sealed, a slow process of absorption and healing takes place, and eventually the vessel is restored to normal.

William Hewson's book, published in 1771, represented a major advance in the understanding of clotting. Previously it had been thought that the red corpuscles merely solidified in a clot, but Hewson showed that plasma contains a substance which he called coagulable lymph, but which later was to be known as fibrinogen. Hewson described how he had been able to delay clotting and to separate this coagulable lymph from the plasma. Unfortunately, this brilliant haematologist died in 1774, at the age of thirty-six, from septicaemia after he cut a finger during a research dissection.

The mechanism of coagulation was studied throughout the nineteenth century. In 1845, fibrin ferment, the primary activator in what is now known to be a most complex process, was extracted. Thirty years later it was shown that fibrinogen is converted to fibrin.

The first indication of the critical role of the blood platelets in clotting was not revealed until 1874, when Sir William Osler, the great Victorian clinician and medical philosopher, noted that the bulk of the pale thrombus formed in blood vessels was composed of tiny fragments of cells from the bone marrow. Although these fragments had first been described by Alexander Donné in 1842, they were not known as platelets until forty years later when Giulio Bizzozero gave them this name.

In 1877, Georges Hayem, a Parisian haematologist and physician, called the platelets the "third element of the blood"—the first two are the red and white cells. The platelets are the frayed edges of giant cells in the bone marrow called megacaryocytes, which shed parts of their cytoplasm into the circulation. These pieces, or platelets, act as a source of conversion substance to change soluble fibrinogen into insoluble fibrin. And it is the platelets which clump together to form a mechanical meshwork to help repair any injured vessel wall.

Sir Almroth Wright, a professor of pathology who served in the British Army and gained fame as a pioneer of immunization, discovered in 1891 that calcium in the blood is essential for the clotting process to take place. This discovery now has great practical importance, for laboratories and blood banks can maintain blood samples and transfusion supplies in a fluid state simply by adding substances which remove the calcium.

Although blood transfusion is itself a therapeutic procedure, it is indivisibly linked with anatomy and physiology. Without knowledge of the physiological factors involved in blood group classification, there could be no successful blood transfusion—nor any organ or tissue transplantation, of which blood transfusion is the oldest and best-established example.

The earliest attempts to transfuse blood were made before blood types were comprehended. Indeed, not even the composition of blood was known, for it was about this time that the earliest microscopists were peering down at dimly seen globules.

The Cornishman Richard Lower, whose work in connection with the circulation of blood and respiratory physiology was well known, was the first man on the transfusion scene. In 1665, he used a crude device, invented by Christopher Wren, to connect the blood vessels of two dogs. Wren had fashioned a makeshift syringe out of a goose quill and a dog's bladder to inject drugs directly into the circulation. Lower used this, together with a silver tube, to join the cervical artery of one dog to the jugular vein of the other. This, and subsequent experiments over the next two years, were done under the aegis of the Royal Society and were recorded in Samuel Pepys's famous diary.

Some of the recipient dogs survived and so on November 23rd, 1667, members of the Royal Society gathered to witness the momentous experiment of transfusing twelve ounces of sheep's blood into the Reverend Arthur Coga, an impoverished clergyman. The procedure went off happily. Fortunately, there could have been no basic incompatibility between the blood groups, although the clergyman had received blood from another species. Pepys commented that "The patient speaks well, saying that he finds himself much better, as a new man—but he is cracked a little in his head."

Blood transfusion received a severe setback, however, early in its career. Jean Baptiste Denis, a French physician, had been duplicating Lower's experiments, but he risked human transfusion four months before the Cornishman. In June 1667, he gave eight ounces of sheep's blood to a feverish young man who had been bled by his own doctors to the point of exhaustion.

All went well and Denis tried transfusion of

Pioneer attempts at blood transfusion were made in the seventeenth century—long before there was any recognition of the need for compatibility between blood types. The earliest experiments, like the one shown here, involved the transfusion of blood from certain animals to man.

blood in a few more cases. Then, in January 1668, Saint Amant, a valet to Madame Sévigné, died during a transfusion, although it seems that the procedure may not have been the cause of his death. Nevertheless, the patient's widow accused Denis of murder. The courts aquitted him, but the medical faculty of Paris and the Chamber of Deputies forbade any more blood transfusions and the procedure was not undertaken again for almost two hundred years.

Towards the end of the nineteenth century, the technique was revived in England, particularly for cases of maternal haemorrhage at childbirth in which, without treatment, mortality was high. Human blood was used and, by 1875, almost four hundred transfusions had been carried out. Then the introduction of saline infusions for shock reduced the urgent need for these rather heroic blood transfusions.

In 1875, Leonard Landois, a German physiologist, found that if blood from two different species of animals were mixed, haemolysis, a rupture of the red cells and escape of the haemoglobin into the plasma, often occurred. In addition, clumping, or agglutination, of the red cells took place. Landois correctly attributed the passing of black urine after many blood transfusions to this cause. His work stimulated interest

in seeking the reason for the inconstant compatibility of different bloods.

At the turn of the century, the Viennese-born pathologist Karl Landsteiner secured a niche in medical history by discovering the main blood group system—the ABO types. He did this largely by a process of elimination by mixing plasma, or serum, from one person, with red cells from other donors to see if agglutination, an indication of incompatibility, occurred. Grouping by agglutination soon became the accepted technique for typing, or matching, bloods prior to transfusion.

From that time, the new science of blood group serology developed explosively. Landsteiner went on to discover, in 1927, the M, N, and P systems. Many more systems were subsequently described and most of these were named after their discoverers, including those of Kell, Duffy, Kidd, Lutheran, Dombrock, and Lewis. Landsteiner, who joined the Rockefeller Institute of Medical Research in New York in 1922, was awarded the 1930 Nobel Prize for physiology and medicine for his work on blood groups.

A decade later, Landsteiner was involved in isolating the Rhesus factor in blood. This was a major discovery, made with Alexander Weiner, a forensic serologist, in the Office of the Chief Medical Examiner of New York. The Rhesus factor is the cause of the serious condition known as haemolytic disease of the newborn in which the infant's blood group becomes incompatible with that of the mother. The Rhesus system takes its name from the breed of monkey that was used for the experiments.

From the days when the ancients saw blood as a simple, irreducible substance—essential to life, but useless when spilled—this vital fluid is now recognized as a complex tissue, carrying oxygen and nutrients to all parts of the body. Separated from the body, it still has a role to play. In cases of injury or disease, whole blood or its components —red cells, white cells, and plasma—can be given as a life-saving treatment.

The word lymph comes from the Latin for spring water, a poetic description of the thin, colourless fluid which is derived from blood plasma. Lymph travels in the lymphatic system, a second and separate circulatory system with a complex of channels almost as extensive as the blood vessels. The lymphatic system has two main functions. The first is to retrieve the lymph from the tissues and bring it back into the main circulation via the thoracic duct in the chest. The other function is the formation of some types of non-granular leucocyte, including the lympho-cytes which are concerned with immunity. The lymphatic system, particularly the lymph nodes, are part of the body's complex reticulo-endo-thelial system, which is deeply involved with all immune processes.

On their way back from the tissue spaces, the lymphatics pass through the lymph nodes, or glands, which act as filters and as production centres for lymphocytes. The lymph glands are clustered at strategic points, including the groin, armpits, and neck, where they filter the lymph returning from the limbs and the head. Other masses lie in the root of the intestine guarding the lacteals, the lymph vessels of the gut.

Bacteria are eliminated by the massed lining cells of the sinuses of the lymph nodes. When infection is virulent, the nodes themselves become inflamed, accounting for the painful swelling that may be felt in the armpit or groin. When inflamed to the point of sepsis, the glands may actually break down into pus. Bubonic plague gets its name from the buboes, or swollen glands, caused by the plague bacillus, which were the horrific hallmark of the Black Death.

Lymphatic vessels were seen by Aristotle, Herophilus, and Erasistratus, all of whom considered them to be veins. Galen denied their existence and, consequently, nothing more was heard about them until the sixteenth century. Vesalius noted the mesenteric lymphatics, but called them veins. Fallopius, in 1561, and Eustachius, in 1564, both mentioned the lym-

phatics. Each described the whitish or yellowish vessels coursing about the body, but their observations were almost incidental and they had no idea of the purpose of the channels.

The lymphatic system actually began to be investigated in the seventeenth century. In 1622, Gasparo Aselli, professor of anatomy and surgery at Pavia, discovered the lacteal vessels. Although these vessels had been observed earlier, it was Aselli who described them after seeing the fat-filled lacteals of a dog's intestines. Aselli thought that the lacteals—whitish cords distended with chyle from the small intestine— were nerves until he saw the rich, creamy fluid which was exuded when one was cut open. In his tract *De Lactibus*, which was published in 1627, the year after his death, Aselli described these vessels as the *venae albae et lacteae*, the white and milky veins.

In 1647, Jean Pecquet, a twenty-five-year-old medical student at Montpellier, dissected the thoracic duct of a dog and showed that it rejoined the main vein of the upper chest, the superior vena cava. This junction was for some time called the canal of Pecquet. He followed his discovery of the thoracic duct by describing the receptaculum chyli, the small pouch which accumulates the fatty chyle coming from the lacteals. This was often called Pecquet's cisterna. He published an account of these structures in 1651 in *Experimenta Nova Anatomica*. Pecquet, who later became physician to Mme. de Sévigné and a distinguished surgeon in Paris, nevertheless continued to be interested in the lymphatics. He investigated the effects of alcohol on the body and since he used himself as a subject this may well have contributed to his death in 1674.

Strangely, William Harvey refused to believe either Aselli's or Pecquet's published accounts, although it seems that what he took exception to was their claim that the lacteals were the sole means of absorbing food products from the gut. Harvey was convinced that the blood also plays a part and, of course, in this he was correct.

In about 1653, Olaf Rudbeck, a Swedish anatomist at the University of Uppsala, discovered the lymphatic vessels, which he called vasa serosa, and noted the clear watery fluid they contained. Thomas Bartholin, the Danish anatomist of the famous Copenhagen medical dynasty, who was undoubtedly aware of Aselli's work and of the investigations of his contemporary Rudbeck, worked out a scheme of the system. He called the vessels vasa lymphatica, because of the watery fluid within them. Bartholin was more eminent an anatomist than Rudbeck and for this simple reason it was his term that was adopted.

Bartholin may not have been familiar with the work and discoveries of Pecquet. It is certainly doubtful that he knew that an Englishman named Jolyffe described the lymphatic vessels in his doctor's thesis at Cambridge in 1652. In 1698, William Cowper, the English surgeon and anatomist, wrote: "The knowledge of this Animal-liquor called Lympha, and the ducts which Convey it, is owing to the Industry and Searches of this present Age. But whether Rudbeck, Bartholine, or our Countrey-man Dr. Jolive ought to carry the Honour of the Discovery, I shall not pretend to decide."

In England, Richard Lower conducted some experiments in 1669 which demonstrated the way lymph re-enters the circulation. William and John Hunter, too, added to knowledge of the system. Then, in 1774, their pupil William Hewson discovered that white cells are produced in the lymph nodes and first noted two sets of lymphatic vessels—the superficial and the deep.

The definitive eighteenth-century work on this complex system was that of Paolo Mascagni, a professor of anatomy at Siena. In 1787, he produced a beautiful book on the lymphatics with copperplate engravings that were for a long time the best illustrations available. The book became a standard work on the subject and appeared in many editions. Like other investigators, Mascagni had seen the migration of leucocytes through the walls of blood vessels in areas of inflammation. This amoeboid movement, although noted, was not understood as the phenomenon of phagocytosis for almost one hundred years.

It was Hewson who had realized that white blood cells originate in the lymph nodes, but in 1885 Walther Flemming, a German anatomist, discovered the germinal centres in the lymph nodes which are the nests where the lymphocytes are produced. Flemming was a pioneer in the field of cell division, and the lymph node centres were the first sites in the body in which cell division was actually observed.

The lymph nodes are part of the body's diffuse reticulo-endothelial system, which is concerned with immunity and defence against infection. It was first proposed as a physiological entity in 1913 by the great German pathologist Ludwig Aschoff.

The reticulo-endothelial system comprises all the lymphatic tissue in the lymph nodes, the spleen, the protective circle of lymphatic tissue in the pharynx, consisting of the tonsils and adenoids (called Waldeyer's ring after Heinrich Waldeyer, a German anatomist), the thymus gland, and the Kupffer cells of the liver, named after Karl Kupffer of Munich who described them in 1876.

Some parts of the system shrink after childhood, particularly much of such lymphoid tissue as the tonsils and adenoids and, most notably, the thymus. The thymus, which in an infant is as large as the heart, is situated in the upper part of the chest. After puberty it almost vanishes and is represented in the adult only by a

Carl Wilhelm von Kupffer was not an original thinker. He saw pioneer works as stepping stones to more thorough research. A native of Baltic Russia, Kupffer was born in 1829. He qualified as a doctor and held the chair of anatomy at Kiel, Königsberg, and Munich. He was a notable embryologist, but had diverse interests and did critical work of an intricate and time-consuming nature in neurology, craniology— he wrote a memoir of Kant's skull —arctic science, and histology. The first to understand the character of protoplasm and to describe the fertilization of an egg, Kupffer continued to work until his death in 1902.

small pad of fat and fibrous tissue. When it is active, the thymus is packed with lymphocytes. A normal cavity in its centre contains liquid composed entirely of these cells. It is called Dubois' abscess, although it is not clear which of the numerous Dubois in medical history is commemorated. A well-known eponym in the thymus, however, concerns the laminated microscopic nodules which are known as Hassal's corpuscles. They are named after Arthur Hill Hassal, a London doctor who, in 1846, published the first English textbook of histology in which he described these curious concentric bodies.

The other major organ of the reticulo-endothelial system is the spleen, a large, soft, red structure that hides under the ribs on the left side of the body. The spleen, which serves as a graveyard for the red blood cells, was well known to the ancient Greeks, who at one period removed it surgically because they thought it improved athletic performance. It was clear from ancient times that removal of the spleen did not affect normal health and it is now known that the spleen's function of destroying red cells is, in such cases, taken over by other organs of the reticulo-endothelial system.

Although Mondino, Jacob Berengar, Estienne, and Vesalius all described the spleen, it was Marcello Malpighi who, in 1659, gave the first adequate description of its structure and of the corpuscle formations which have since been irrevocably known as Malpighian corpuscles.

Today, the anatomy and the microscopy of the reticulo-endothelial system has been investigated. Yet understanding of its complex physiology in relation to the body's immune system continues to be fertile ground for researchers. Each new discovery seems to open the way for yet another. It is obvious that much still remains to be learned about the almost limitless immunological potential of the human body.

The Cell

During development in the womb, cells differentiate into specialized tissues. Oblong cells, striped in appearance and endlessly branching, characterize the muscle of the heart.

The human adult body is made up of several hundred million million cells. Of these there are thousands of different types of cells, each with its own place and function in the body. Although a cell is usually thought of as spherical, actual functioning cells in tissues and organs take on a variety of shapes. The cells of the liver, for example, are hexagonal. Those which line the intestines are column-shaped, and muscle cells are elongated into fibres.

The fundamental unit of both plant and animal life, the cell has been recognized as a vital entity for less than one hundred and fifty years, although the visible cellular structure of tissues had been known for much longer. The study of the cell was entirely dependent upon a means of magnifying the tissues sufficiently, for, with the exception of some egg cells, the ultimate units of life are below the limits of human vision. Each cell is a microscopically tiny unit surrounded by the cell wall, a delicate membrane about sixty-millionths of an inch thick.

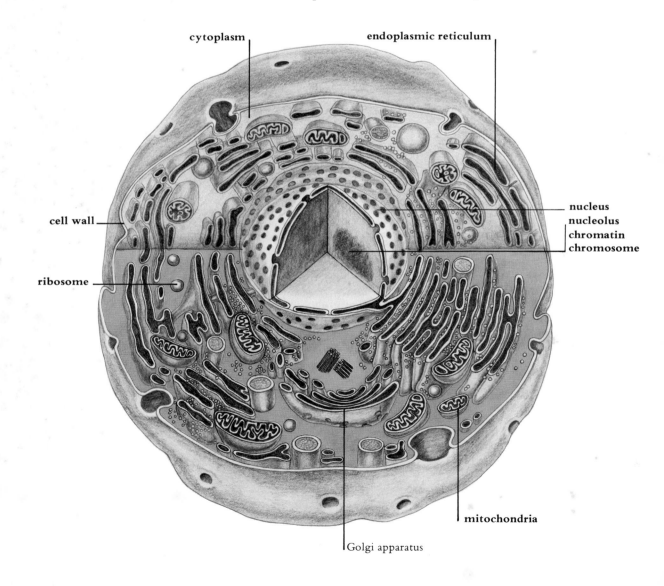

cytoplasm

endoplasmic reticulum

cell wall

ribosome

nucleus
nucleolus
chromatin
chromosome

mitochondria

Golgi apparatus

Through the cell wall oxygen passes into the cell, dissolved in the fluid that bathes every tissue. Molecules of certain food materials are also absorbed. At the same time, carbon dioxide and molecules of waste material, the result of the cell's unceasing activity, pass outwards through the cell wall and are collected by the bloodstream to be ultimately disposed of by the body's excretory system.

The reproduction of each cell is directed by its nucleus, a spherical structure containing the chromosomes, the basic genetic material. Within the nucleus is the nucleolus, which is concerned with the formation of nucleic acids, necessary for the production of the protein material which make it possible for the cell to grow, divide, and repair itself. The bulk of the cell is formed by cytoplasm, a clear fluid, resembling the white of an egg, in which all the structures are suspended.

In the days before the microscope, questions about the nature of living matter were unanswerable. What speculation there was tended to the view that animal tissues were perhaps a continuous slime or possibly a spongy matrix.

Blood corpuscles were among the first human cells to be observed by such early microscopists as Anton van Leeuwenhoek. The honour of describing the configuration of a cell and of actually using the name for the first time, goes, however, to Robert Hooke, one of the remarkable group of men in Restoration England who founded the Royal Society, the premier institution then and now for the promotion of scientific knowledge.

Hooke was the society's curator of instruments and in 1663 he did some research which was published the following year in a book on wood entitled *Sylva* and written largely by John Evelyn. Hooke described his examination of thin slivers of petrified wood in which he observed "microscopical pores." Two years later, his own book *Micrographia* was published and in this he amplified his work on wood, among other things, and described the appearance of slices of cork, which he studied under his microscope. He found that it was made up of little boxes or compartments and described his sections as "all cellular or porous in the manner of a honeycomb, but not so regular."

Hooke saw, too, that when fresh vegetable tissue was examined, the cells usually contained fluid. He wondered whether there was any communication between one cell and the next. Although he could detect none, he cautiously did not assume that there was none.

In 1670, Marcello Malpighi published *Anatomy of Plants,* which contained a description of cells that was more accurate, and was considered more significant, than Hooke's work.

In 1682, Nehemiah Grew, another curator of

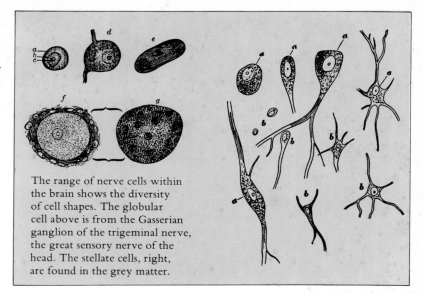

The range of nerve cells within the brain shows the diversity of cell shapes. The globular cell above is from the Gasserian ganglion of the trigeminal nerve, the great sensory nerve of the head. The stellate cells, right, are found in the grey matter.

the Royal Society, also wrote an *Anatomy of Plants*. He described the structures seen by Hooke, but called them "pores" and "bladders" rather than cells.

Other microscopists made their own observations and described cells as morphological entities. But they seemed to have little curiosity about the function and status of these tiny structures in the basic life process. Understanding of the fundamental nature of the cell lay dormant for almost two centuries, until the earlier years of the nineteenth century.

The idea that the whole organism is built up of an aggregate of small cell units was not contemplated. The cell walls, obvious in plants but, with the poor instruments then available, imperceptible in animal tissues, were believed to be a fibrous framework. The fluid within the walls was not thought of as being in separate cells, each cell able to live a relatively independent life—or, at least, able to live as a single entity within a huge colony.

In 1781, an important discovery for future scientists was made by Felix Fontana. A professor of anatomy at Pisa, Fontana was the first to note the nucleus in an animal cell. In a book about snakes, Fontana described seeing "ovoid bodies" inside the epithelial cells of the skin of an eel. He also mentioned a spot within the nucleus—no mean feat considering the primitive microscopes of his day—which might have been the nucleolus.

Fifty years passed before the nucleus was seen within a plant cell and it was recognized that a nucleated cell is the fundamental unit of all vegetable tissues.

By the beginning of the nineteenth century, cytology, the study of cells, was becoming a popular occupation among botanists and, to a lesser extent, zoologists. This was mainly due to the improvement in optical instruments, particularly the development of lenses which corrected spherical and chromatic errors. Once the resolving power of microscopes was improved, the way was open for more detailed observations of the tissues and more deductions from these observations.

In 1805, for the first time vegetable cells were teased apart and shown to be individual units. Nineteen years later, René Dutrochet delivered a paper at the Paris Academy. In it he said, "... we have seen that plants are composed entirely of cells, or of organs that are obviously derived from cells; we have seen that these cells are merely contiguous and adherent to each other by cohesion, but that they do not form a tissue exactly continuous.... All of the organic tissues of plants are made of cells, and observation has now demonstrated to us that the same is true of animals." This statement had considerable clarity, but it went unheeded for many years, although gradually the unit nature of the cell began to be appreciated.

By the 1830s, the time was ripe for a major advance in recognizing the all-important role of the cell in living organisms. Both sides of the biological duet of botanist and zoologist realized that the cell is the basic unit in each of the two kinds of life types—the plant and the animal. Two men stand out in this context— the German botanist Matthias Jakob Schleiden and Theodor Schwann, the brilliant German anatomist, whose name is commemorated in the sheath of the nerve fibre.

Schleiden was an unusual character. He had been a barrister in Berlin, but disillusioned by his failure to succeed in this profession, he attempted suicide. When he recovered he gave up the law and began a new career. He studied science and became a professor of botany.

In 1838, Schleiden announced that the cell is the basic structure unit of all vegetable matter, the first time that this concept had been unequivocally stated. Schleiden had a strong influence on young men in the German scientific world, which at that time was taking over the lead from France in laboratory science. Theodor Schwann was one of these men, as was Carl Zeiss, who was to found the world's most famous optical works at Jena, where some of the best microscopes ever made were to be produced.

Schwann dated his inspiration and all his interest in the cell to one occasion when he dined with Schleiden. He came away strongly impressed by the concept of the cell as fundamental to animal tissues, as well as to plants. The theory that the cell with its nucleus is the ultimate building block of all living tissues was reinforced by a recollection that he had seen cells similar to Schleiden's vegetable units in the cartilaginous spines of young fish. Schwann decided that animals and plants shared this same dependence upon cells and declared that "cells are organisms and entire animals and plants are aggregates of these organisms arranged according to definite laws."

Like Harvey's theory of circulation of the blood this part of Schleiden's cell theory seems patently obvious to us today. Early in the nineteenth century, however, it was a bold statement and one that took a considerable time to become accepted.

Unfortunately, although his basic concept of the ubiquity of cells was correct, the rest of Schleiden's theory was wrong, a mistake that set back the progress of cytology for about fifty years. Schleiden's major misapprehension concerned the development of cells, which, of course, always arise from a parent cell by means of splitting, or binary fission.

Schleiden and Schwann both believed, however, that cells arose directly from the body fluids or that the nucleus of an existing cell was a kind of embryo which produced a new cell. The idea evolved that the future cell membrane was the nuclear membrane which enlarged until the new cell matured at the expense of the parent. The other mechanism was thought by Schleiden to arise from the general tissue fluids, called the blastema, in which first a nucleolus—the small central core of the nucleus —appeared, then surrounding granules condensed to form the nucleus, followed by further condensation to produce the rest of the cell.

This was, in effect, a resurrection of the old concept of spontaneous generation, although it occurred internally within the body of the plant or animal. This theory of free cell generation persisted until late in the nineteenth century when it was demolished by Rudolf Virchow, the great pathologist and Father of Cellular Pathology.

In 1839, Johannes Purkinje, the Bohemian

Matthias Schleiden was born in Germany in 1804. Acclaimed as the discoverer of the universality of the cell in plant structure, he studied law and medicine before he became a botanist. A professor of botany at Jena, Dorpat, and Frankfurt, Schleiden rejected the current emphasis on the study of plants by their visible features; he preferred to examine plant structures under the microscope. In 1838, he published a paper entitled "Phytogenesis," in which he proved that plant tissues are made up of cells. One of the first German scientists to support Darwin, Schleiden died in 1881.

physiologist, coined the word protoplasm for the cell contents, although his meaning was slightly different from the modern usage. Theologians had long called Adam the protoplast, meaning the first-formed, and Purkinje used the analogy in the cell for the first-formed substance when the cell was made. He wanted to convey that the gelatinous ground substance, which Nehemiah Grew had long before called the cambium, was the original substance of each individual cell.

That the protoplasm was not a static jelly was first observed by the Abbé Corti in 1774, when he saw streaming movements inside the cells of water plants. This was forgotten until about 1811. It was then thought, however, to be a phenomenon confined to large plant cells; this "cyclosis" could be more easily seen in them because of the large chlorophyll-containing inclusions which rotate within the cell sap. Later it was observed in other cells, including blood leucocytes which have amoeboid movements.

In 1861, Max Schultz, the German anatomist who had studied at Berlin with Johannes Müller, defined a cell as a mass of protoplasm containing a nucleus and identical in plants and animals. One year later, Rudolf von Kölliker, the Swiss anatomist, first used the term cytoplasm, from the Greek words for hollow (or a cell) and something formed or moulded, as the equivalent of protoplasm. Later, its use was restricted to mean the substance of the cell as distinct from that of the nucleus.

But no one yet understood how new cells came into being. The erroneous theories of Schleiden and Schwann persisted until three-quarters of the way through the nineteenth century. Various odd descriptions were put forward in efforts to perceive stages in development. Many of them concerned the "primordial utricle," meaning a layer of protoplasm that lined the thick wall of plant cells. Gradually, Schleiden's theory was destroyed and it was shown by direct observation of algae that some cells could split into two daughter cells.

Carl Nägele, one of Schleiden's protegés, turned from defence of his master's hypothesis to declaring that cells divided by splitting everything into two halves, although he allowed that some cells could obey the Schleiden principle.

No further progress could be made until it was appreciated that only a nucleus could make another nucleus and a cell another cell. Work on developing egg cells was proceeding—indeed many had been watched for years before, going back to the time of Jan Swammerdam in the seventeenth century. These earlier observers did not grasp what was happening, however. It was another century before Karl von Baer, the Estonian zoologist, saw the nucleus splitting during the cleavage of an egg cell.

Johannes Evangelista Purkinje, the Bohemian physiologist, was one of the great figures in the developing science of histology in the nineteenth century. Born in Prague in 1789, he was professor of physiology at Breslau and later at Prague. In 1842, Purkinje founded one of the first laboratories of physiology. He introduced new techniques in microscopy and made many notable discoveries. He investigated vertigo, deafness, ciliary movement, digestion, and noted that each person has unique fingerprints, the importance of which was not realized for almost fifty years. Purkinje died in Prague in 1869.

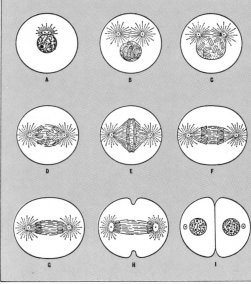

The replication of cells by division is a process, usually lasting about an hour, known as mitosis. As the division begins, chromosomes containing the basic material of heredity become shorter and fatter. The nuclear membrane distintegrates, freeing the chromosomes to replicate and to attach themselves to a meshwork forming in the cytoplasm, which is the fluid medium of the cell. Next, the chromosome halves are drawn to opposite poles of the cell. The cytoplasm divides and two nuclear membranes reform.

Robert Remak, a German neurologist and a professor at Posen, made a large stride forward when, in 1852, he watched the embryonic blood cells of a chick dividing and realized what was happening. He concluded that in normal conditions a cell always divides into two identical daughter cells, with the nucleus splitting prior to the separation.

A friend of Remak's, Rudolf Virchow, was one of the greatest names in nineteenth-century medicine and one of the greatest pathologists of any century. Although Virchow had accepted the Schleiden-Schwann theory at the outset, he came to reject it and to destroy it completely by the time he finished his work on cellular development. In 1855, Virchow coined the well-known aphorism *Omnia cellula e cellula*—every cell from a cell. A political radical who manned the revolutionary barricades in

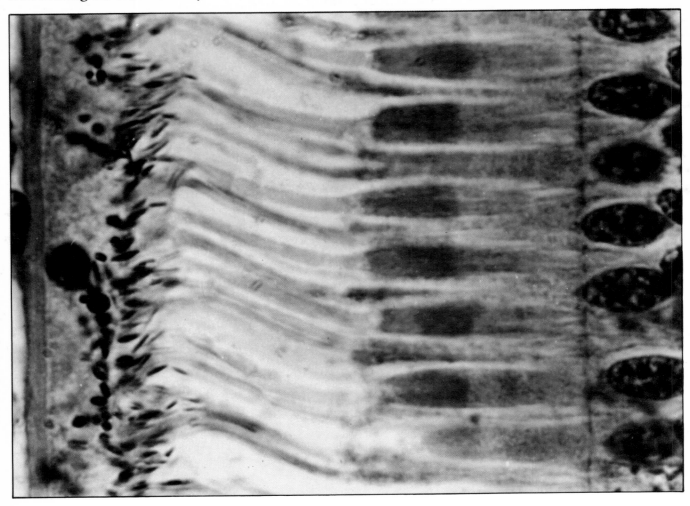

The light-sensitive cells of the retina—seen above magnified seven hundred and eighty times—are known as rods and cones because of their distinctive shapes. The thin cylindrical cells are the rods, concerned with black and white vision. There are about 125 million of them in the human eye. Bulbous in shape, the cones, about five million in each eye, transact colour vision.

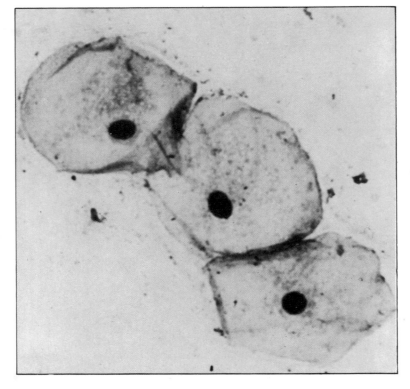

Epithelial cells are those whose function is to cover surfaces inside the body or out. These include skin cells, which are layered and become flattened the closer to the surface they are. At a magnification of one thousand times, the nuclei of these cells from the lining of the cheek are clearly visible.

Germany, Virchow saw the cells as members of a socialist state, each one working towards the common goal. In his famous book on cellular pathology, published in 1858, he described the growth and division of cells, named and described many pathological conditions, and introduced a new concept of thrombosis and embolism. It was a milestone in the understanding of the cell in health and disease and it remains one of the most notable medical textbooks of all time.

The problem of nuclear division was the key to further understanding and much was contributed by an amateur biologist, rather in the mould of the linen draper Anton van Leeuwenhoek. W. Hofmeister was a German bookseller who was inspired by Schleiden's book to take up microscopy. Every day he would devote two hours, from four o'clock to six o'clock in the morning, to his hobby, before going off to work in his father's business in Leipzig. Hofmeister carried out painstaking work on plant cells and saw that the nuclear membrane dissolved before the cells divided, although the nuclear contents were still visible. The nuclear mass then split into two and a cell membrane formed between the two halves to produce the daughter cells.

No better observations were made for another thirty years after Hofmeister's publications on the subject. He was eventually given an honorary degree by the University of Rostock and later made a professor of botany first at Heidelberg and then at Tübingen.

The whole process of cell division now centred on the nucleus and the way it divided. Remak had noticed a thin band running between two nuclei in a blood cell that was dividing. Other microscopists had seen a radiating appearance in egg cells just before division, which is now known to be the separating process of the chromosomes.

In the middle of the nineteenth century, it was first observed that there is an astral pattern in cells, with radiating threads pulling the split chromosomes into each half of the dividing cell. Most of the knowledge accumulated until 1882 was consolidated by the German anatomist Walther Flemming, who added a substantial amount of his own work in a monograph he published in that year. A professor of anatomy at Kiel, Flemming gave a good account of the way in which the nucleus divides, although he thought that the unravelled threads of chromatin, the nuclear stuff of heredity, were one long strand, rather than separate pieces, as is known today. The number of pieces equals the number of chromosomes.

The word chromosome was introduced in 1888 by Heinrich Waldeyer, professor of pathological anatomy in Breslau and then at Berlin. It is derived from a Greek word meaning coloured bodies and refers to the stainable nature of the nucleo-protein material when prepared by various dyes before microscopic examination.

The process of nuclear splitting, in which each chromosome divides longitudinally and is then pulled into each daughter cell, was named mitosis by Flemming. The word is derived from the Greek for threads, because of the appearance of the chromatin substance just before division.

The human being has forty-six threads, or chromosomes, in each nucleus, which carry all the genetic information for the building of every body tissue and organ. That the human is the most complex animal in existence is not reflected in the number of chromosomes, for a relatively simple one-celled organism, a rhizopod, has fifteen thousand chromosomes.

When sex cells are formed, as in the testicular formation of sperm or of the egg in the ovary, the cells require only half the number of chromosomes, because they are going to combine with the sex cells of the other partner in order to reform the forty-six total. The process of halving of the sex cells was named by J. B. Farmer and J. E. Moore in 1905, who used the Greek word maiosis, meaning to lessen. This became altered by later usage to meiosis, a term now universally recognized in the biological sense, although it is not quite accurate etymologically.

The story of the way inheritance within the cell system was worked out is complex and vast, but a few highlights must be mentioned, and a few names recorded. The first, of course, was Mendel, whose name has been perpetuated in the most lasting of eponyms, Mendelism. Gregor Mendel was born in 1822 in the village of Heinzendorf in what is now Czechoslovakia. He was the son of a peasant who was experienced in fruit-growing, a fact which became relevant in shaping the son's latest interests.

Karl Ernst von Baer, who is called the Father of Modern Embryology, was born in Estonia in 1792. He was professor of zoology successively at Dorpat, Königsberg, and St. Petersburg. In 1827 he published *On the Mammalian Egg and the Origin of Man.* In the later years of his life, von Baer devoted himself to anthropology and, independently of Darwin, suggested a common ancestry for diverse species. Von Baer's philosophy ruled his science, however, and he rejected evolution and the possibility of human kinship with animals. Before he died, in 1876, he founded anthropological societies in Germany and in Russia.

Gregor Mendel, the monk from Czechoslovakia who lived from 1822 to 1884, laid the foundations of modern genetics. Mendel studied for two years before he joined the Augustinians at Brünn, but he failed the teachers' examinations in biology and geology. His abbot, however, sent him to university to study science, including mathematics and botany, and when Mendel returned to Brünn, he began to teach and resumed his own research. The results of his years of work, *Experiments with Plant Hybrids,* were published in 1866. The importance of his conclusions was not, however, appreciated during Mendel's lifetime.

Camillo Golgi devised the silver nitrate stain—used to give better definition of nerve tissues under the microscope— in 1873 when he was thirty and still a physician at a home for incurables near Pavia. His work on cells soon brought him recognition, and he became professor first of histology and later of pathology at the University of Pavia. A brilliant neurohistologist, in 1883 he wrote "On the Fine Anatomy of the Nervous System," for which he was awarded the Nobel Prize in Medicine in 1906. In 1885, Golgi embarked on research to identify the parasites implicated in different types of malaria. He died in 1926.

Gregor entered an Augustine monastery and later rose to become its abbot, but in his earlier years, he devoted much of his time to gardening and cross-pollinating vegetables, especially peas, upon which almost all his work on heredity was carried out. In his plot in the Augustinian monastery, he grew twenty-two varieties of peas and became fascinated by experiments in cross-breeding them.

He made a relatively simple, but vastly important discovery which has become the cornerstone of the laws of heredity. He discovered that when tall and short peas are crossed, or hybridized, the first generation all grow tall, because this trait is dominant. But of the second generation, one-quarter breed true short peas, one-quarter breed true tall peas, and the remaining half are identical with the parent plants, being tall but with variable breeding properties.

Although Mendel did not concern himself with the cell or connect his work with the contents of the nucleus, the laws he discovered are directly consequential on the chromosome structure. In 1868, when Mendel was appointed abbot of the monastery he abandoned his studies in natural history. His work was unknown to most of the scientific world when he published it in 1865. Little regard was paid to it until the beginning of the twentieth century, when it was rediscovered and applied with considerable excitement to the genetic revolution that was emerging from cytology.

The basic mystery was how the filamentous threads of the cell nucleus carry coded information which allows the transmission of genetic characteristics. It was known that the chromosomes split lengthwise during mitosis and meiosis, but what magic within them passed the messages from generation to generation?

The first step was to determine what the chromosomes were made of. This work was begun in the late 1860s by Frederick Miescher of Basel, whose father had been a pupil of the famous Johannes Müller and who himself had studied under Wilhelm His. Miescher decided to make chemical analyses of nuclei that were separated from the rest of the cell protoplasm. As a source of abundant nuclei he used the discarded bandages from the hospital, which were often soaked in pus that is composed of untold millions of leucocytes from the blood. This was an offensive source of research material, but a scientist hot on the trail of research is rarely put off by such aesthetic considerations, although Miescher admitted to rejecting some of the most nauseating material.

Miescher's efforts were rewarded by the isolation, after long and difficult analyses, of an acid substance rich in phosphorous. His director in the research was Ernest Hoppe-Selyer, the noted German physiological chemist. He was so impressed by the findings that he repeated all the experiments himself before permitting Miescher to publish his findings in 1871. Miescher called the acid substance nuclein but, after further work on fish sperm, began calling the new compound nucleic acid, which he discovered combined with a protein that he named protamine. Later work by other biochemists showed that Miescher's nuclein was actually a group of compounds, and eventually the vitally important deoxyribonucleic acid (DNA) and ribonucleic acid (RNA) were isolated.

In about 1915, it was realized that chromosomes are not uniform structures, but are like strings of beads. The individual beads were called genes, derived from the Greek word for race. With only forty-six chromosomes in humans, it was obvious that there are insufficient numbers in the permutations of genetic information needed to build a human body from a single fertilized egg. The discovery of genes—up to one thousand two hundred and fifty on each chromosome—gave more than enough variables. Whereas there are about eight million ways in which the paternal and maternal chromosomes can combine, the statistics of gene variability are almost beyond comprehension. The possibility of two human beings having the same gene structure by chance is more than one in 10^{9000}.

The last great investigation into the nature of the genes was one of the most famous stories of modern scientific research. It culminated in the award of the Nobel Prize to James Watson and Francis Crick for their work published in 1953. This was the well-known double helix hypothesis, based on crystallography and radiography of the nuclear contents. The various components of DNA were already known, but no one knew how they were arranged in the chromatin material.

Crick and Watson carefully built a three-dimensional pattern of the chromatin thread, discovering it to be a spiral helix like a twisted ladder. The side members were made of sugars and phosphates, the rungs were combinations of four nitrogenous bases, comprising adenine, thymine, cytosine, and guanine. The number of permutations is vast, but the fascinating thing is that the base combinations are fixed, so that when the helix splits longitudinally, each half of the ladder reconstitutes itself into an identical copy of the original, ensuring that exactly the same chromatin pattern passes into each daughter cell. These are the actual coding elements for the genetic information.

In the human cell, the material would be about three feet long if it were placed end to end and the number of rungs reaches about six thousand million. The possible permutations of these give a number which is too large to be comprehended, all of it available for handing on as a hereditary template to the new cells.

Crick and Watson, working at the Cavendish Laboratory at Cambridge, published their famous paper in *Nature* on May 30th, 1953. This announcement to the scientific world was one of the most epoch-making discoveries in the whole history of science, opening so many new avenues, the latest being so-called genetic engineering, in which the chromatin structure of cells can be deliberately altered.

The internal structure of the cell had to await the electron microscope before any idea of its appearance and operation could be gained. The zenith of optical investigation was the beginning of the twentieth century and progress in understanding the ultrastructure of the cell marked time until a new technology

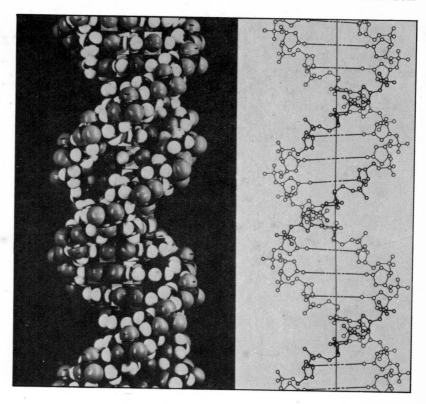

arrived. This came in the late 1930s and in the years after World War Two, in the shape of the electron microscope, followed some years later by the scanning electron microscope, which was able to reveal external surface features of cells at enormous magnifications.

Both the EM and the SEM have revolutionized knowledge of the interior of the cell which has been found to be far more complex than light microscopy ever suggested.

The major microanatomical features clarified by the fantastic magnifications and high degree of resolution of these new instruments have been the endoplasmic reticulum, the network of flattened sacs and tubes which provide a communication channel for materials passing between the nucleus and the cell's environment, and the complex structure of the mitochondria, the main sites of energy produced within the cell. Also confirmed has been the Golgi apparatus, the network of threads in the protoplasm discovered in 1896 by Camillo Golgi, who introduced the method of using silver nitrate to stain nerve cells.

As in all branches of science, new discoveries occur in a stepwise fashion dependent upon the introduction of new investigative techniques. Histochemistry was another major tool in the investigation of the cell, providing methods by which structures can be recognized not only by their shape, but by the function they perform when made to carry out certain chemical reactions. Newer techniques are still evolving and these will be the history of tomorrow.

Chromosomes are beaded in appearance, with each bead representing a gene—a length of basic genetic material called deoxyribonucleic acid, or DNA. Popularly known as the code of life, DNA is the substance containing the blueprint for all development in the body from the moment of conception throughout life. The model of the molecular structure of DNA, above left, is the famous double helix. The diagrammatic form, above right, shows the composition of sugars and phosphates, bases and amino acids, which are the constituents of DNA.

The Microscope

While visiting Galileo in Florence, in 1614, a friend looked through an instrument in Galileo's possession and said he saw flies as big as sheep. Hyperbolical, perhaps, but nonetheless with one amazed glance the long journey had begun into the strange world of abstract shapes and structures that is the inside of the human cell.

This first primitive microscope had been constructed in about 1603 by Zacharias Jannsen, a Dutch spectacle maker, and was given to Galileo as a gift. Galileo introduced the instrument to colleagues at Academia dei Lincei—the Academy of the Lynx—so named because of the lynx's supposed piercing sight. Johannes Faber, a member of the academy, claimed the distinction of inventing the word "microscope" in a work published in 1628.

All the early instruments were light microscopes. They worked by reflecting a beam of light off a mirror, concentrating the beam by sending it through a set of lenses, and passing it through the image being observed. Another set of lenses, focussed into an eyepiece, magnified the image. This process is essentially an amplification of the way the eye sees, by assembling light waves into images.

Malpighi, in 1663, made the first genuinely scientific observations with the microscope when he observed the capillary circulation. In 1665, Robert Hooke published the first work devoted entirely to observations made with a microscope. It was entitled, not surprisingly, *Micrographia*.

Most of the early microscopes were compound; that is, they used several sets of lenses. Anton van Leeuwenhoek, however, was such a skilled lens grinder that his simple microscope with a perfectly ground single lens was better than any compound instrument then available. Neither the inventor nor the first user of the microscope, Leeuwenhoek did such thorough, diligent, and varied work that his name is almost automatically associated with any mention of early microscopy.

Technical developments kept pace with a growing investigative passion, and some of the greatest problems were solved quite quickly. Spherical and chromatic aberrations were largely eliminated in the eighteenth century, and many other improvements followed in the course of the nineteenth century. The inner world of the human cell became progressively and dramatically more visible. In the twentieth century, with the electron microscope, the molecule itself yielded up its structure.

Leeuwenhoek's microscope, below, used a single lens and worked by simple magnification. This is only one of the many types he made, depending on the object to be examined. Indefatigably, he made his observations for most of his ninety-one years and never ceased to wonder at the variety of phenomena he saw.

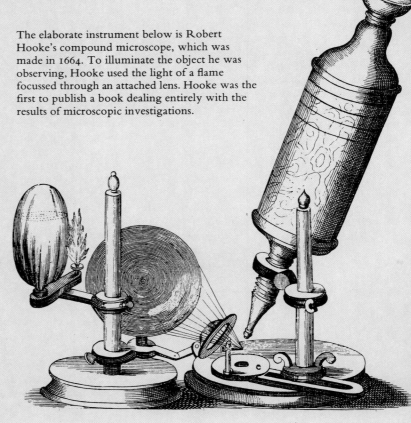

The elaborate instrument below is Robert Hooke's compound microscope, which was made in 1664. To illuminate the object he was observing, Hooke used the light of a flame focussed through an attached lens. Hooke was the first to publish a book dealing entirely with the results of microscopic investigations.

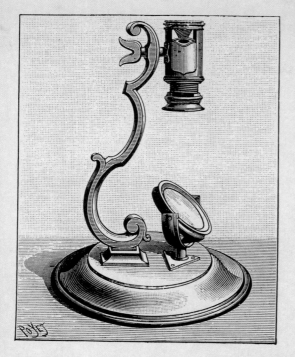

The microscope on the left was made in 1745. A simple instrument, it used a single lens and was essentially a mounted magnifying glass. The simple, hand-held microscope right, had a screen to protect the eye from the glare of light and a rather elaborate attached fixture for the object being examined, in this case a leaf. The microscope below left dates from 1874 and is compound, using two lenses or sets of lenses. The lens attached to the instrument focusses light on the microscope stage while the other lens magnifies the image.

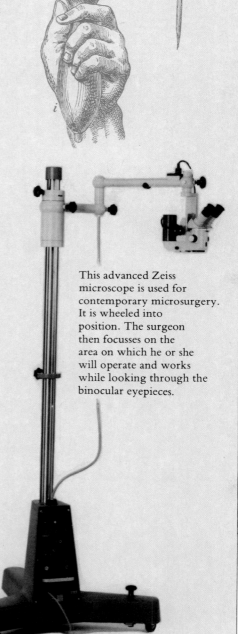

This advanced Zeiss microscope is used for contemporary microsurgery. It is wheeled into position. The surgeon then focusses on the area on which he or she will operate and works while looking through the binocular eyepieces.

The Endocrine System

In the seventh book of *De Fabrica*, Vesalius illustrated the pituitary, marked A, and described it as "the gland which receives the phlegm from the brain."

The newest branch of physiology, endocrinology had its origin almost entirely in the nineteenth century. Even its name is new. The word endocrinology is derived from two Greek words meaning separated within, a rather obscure allusion to the mechanism whereby chemical messengers circulate within the body, stimulating specific and separate effects. A more comprehensible meaning attaches to the word hormone, the name for the individual chemical messengers. This word was coined in 1905 by the English physiologist E. H. Starling and it derives from the Greek to excite.

Most of the "exciting" in the body is done by the nervous system, although part of the transfer of nerve impulses and of the stimulation of muscle by nerve is accomplished through chemical mediators. Many other communications are carried out, however, by an invisible process that requires neither nerve fibres nor junction boxes but, instead, uses the general body circulation for its medium of dissemination. In many ways,

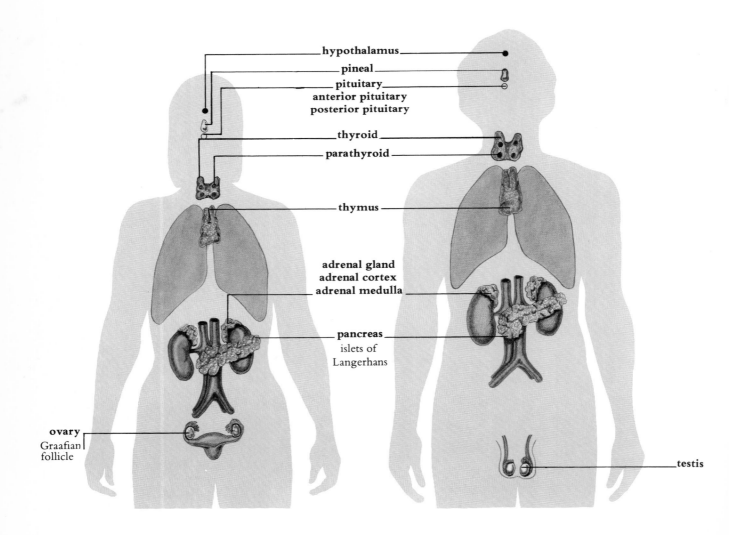

hypothalamus

pineal

pituitary
anterior pituitary
posterior pituitary

thyroid

parathyroid

thymus

adrenal gland
adrenal cortex
adrenal medulla

pancreas
islets of
Langerhans

ovary
Graafian
follicle

testis

the two systems might be compared to the telephone and to the radio. The nervous system needs cables to send its messages. The endocrine system broadcasts its messages generally, although only certain specific tissues are tuned to the correct frequency to receive them.

The actual transmitters, the endocrine glands, have been known to anatomists for centuries, but their functions were not even guessed at until relatively recently. The endocrines were long known as ductless glands, and that is exactly what they are, for they pour their chemicals into the bloodstream, rather than down a pipe or duct, as do the salivary and sweat glands, which are among the body's exocrine glands.

In the middle of the nineteenth century, the French physiologist Claude Bernard put forward his concept of the *milieu intérieur*, the stable internal environment of the animal body, regulated by a multitude of interacting control mechanisms geared to maintain the body's chemical and physical status quo.

Bernard postulated the theory that internal secretions, as part of the body's regulatory mechanism, maintain equilibrium in response to a range of altering conditions imposed on the body. This mechanism, called homeostasis, is essentially that which is accepted today, although Bernard had no means of knowing the details of the way the ductless glands actually operate.

It was not until the twentieth century that a hormone was first isolated. This was the substance secreted from the intestine which stimulates the pancreas to produce its digestive juices. It is released into the bloodstream when food enters the duodenum from the stomach. William Bayliss, the English physician, and Starling isolated it in 1902 and called it secretin. The discovery of secretin stimulated the search for other hormones—a search which still continues.

The major endocrine glands are the hypothalamus and the pituitary at the base of the brain, the thyroid and parathyroid glands at the front of the neck, the adrenals, which are above the kidneys, the pancreas, which lies in the loop of the duodenum, and the ovaries and the testes. In addition, other organs also produce hormones. Several in the gastro-intestinal tract, for example, are involved in controlling digestion.

Some glands are both exocrine and endocrine. The testes, for example, produce both seminal fluid and the hormone testosterone. Similarly, the pancreas has an exocrine function for digestion and an endocrine role for the production of insulin. Other glands, including the pituitary and the adrenals, produce more than one hormone.

The hypothalamus, which is situated at the centre of the underside of the brain, below the thalamus and above the pituitary, is connected to the brain and spinal cord by many nerve tracts.

This tiny structure acts as the link between the endocrine system and the nervous system. It functions automatically, monitoring and regulating the autonomic nervous system as well as the state of the body's metabolism as the result of eating, drinking, temperature control, sexual drives, and fight or flight reactions. It controls the menstrual cycle in a woman and is believed to contain a pleasure centre. It also directs hormones released by the anterior pituitary gland.

The pituitary, often called the master endocrine gland, is about the size of a pea. A compound gland, it is probably the most complex of all the endocrines. It is actually two glands, the anterior and the posterior, each of which has a separate function. The posterior pituitary is directly joined to the brain by a stalk and its substance is more like nervous tissue than that of the front section. It was long believed to be the source of nasal mucous, hence its name which means phlegm.

Aristotle thought that the brain secreted pituita, or mucous, and discharged it from the nose as part of the cooling process which he believed to be the function of the brain. In 1524, Jacob Berengar noted the presence of the pituitary body. Nineteen years later, in *De Fabrica*, Vesalius referred to the gland as *glandula pituitam cerebri excipiens*, and declared that it secreted mucous into the nose. Although this idea was finally disproved in the seventeenth century, the name persisted.

Harvey Cushing, the brilliant American neurological surgeon, said that "the pituitary is the conductor of the endocrine orchestra," indicating that this little gland regulates the activities of most of the other glands in the body, particularly the sexual organs which in both male and female are themselves powerful endocrine structures. It is their arousal by the pituitary at

Ernest Henry Starling, the English physiologist, was born in 1866. He graduated from Guy's Hospital where he later became professor of physiology. In 1893, Starling's sister married William Bayliss and Starling and Bayliss formed a lifelong friendship and working alliance. In 1899, Starling was elected to the Jodrell Chair of Physiology at University College, London, a post he held until 1923, four years before his death. In collaboration with Bayliss, Starling demonstrated peristalsis. He studied kidney function, lymph and other body fluids, and formulated Starling's law of contraction of the heart.

William Bayliss was born in 1866. He studied medicine at Oxford and at University College, London where, in 1912, a chair of general physiology was created for him, which he held until his death in 1924. Bayliss worked with Ernest Starling, his friend and brother-in-law, on the nervous control of blood pressure, on peristaltic wave movement, and on the biochemical stimulation of an organ. Together, too, they discovered secretin. Bayliss himself isolated the digestive enzyme trypsin. During World War One, Bayliss treated shock experimentally with saline injections, and saved many lives. He was knighted in 1922.

the time of puberty that produces the secondary sexual characteristics.

Until the twentieth century, it was believed that the pituitary was absolutely essential for life, but Cushing and the Romanian physiologist Nicolas Paulescu, who did outstanding work on the pituitary just after World War One, found that with proper precautions a patient can survive after the gland has been removed or destroyed.

A number of different hormones are manufactured by the pituitary. The anterior pituitary gland produces the growth hormone (GH). Insufficiency of this hormone in infancy leads to dwarfism. Overproduction results in acromegaly, when massive growth of the hands and a coarsening of the face disfigure the victim. The gland was not associated with these disorders until late in the nineteenth century and the hormone was not isolated until 1944. Today infantile dwarfism can be treated by the injection of this hormone prepared from human pituitaries removed at post-mortem examinations.

The posterior pituitary gland produces anti-diuretic hormone (ADAH), which promotes water retention by the kidneys. A deficiency in this hormone leads to diabetes insipidus, the passing of great quantities of weak, clear urine. It also makes oxytocin, which raises blood pressure and contracts the pregnant uterus. This was not discovered until the last years of the nineteenth century. Oxytocin has important applications in childbirth to increase uterine contractions and reduce later bleeding.

The thyroid gland at the front of the neck is stimulated by thyroid-stimulating hormone (TSH) from the pituitary gland to produce two hormones—thyroxine and triiodothyronine—which are concerned with the stabilization of the body's metabolic rate. The thyroid gland cannot function properly if iodine is lacking.

Although the thyroid gland was recognized in ancient times, its function was not even considered until the nineteenth century. Removal of a diseased thyroid gland was recommended by the Roman writer Celsus in the first century A.D., but the actual operation is believed to date from the eleventh century when the Arabian physician Albucasis first attempted it.

The word thyroid derives from the Greek word for door, and originally meant a large stone placed against a door to keep it shut. Later the same word was applied to a long shield carried by Minoan soldiers which covered them from neck to ankle and had a deep notch at the top for the chin. It was this shield that suggested the name thyroid to Galen for the cartilage in the neck. The thyroid gland took its name from the cartilage, since the gland itself has no resemblance to a shield. It was first known as a laryngeal gland; the name thyroid gland was given to it in 1646 by Thomas Wharton, the English physician and anatomist.

Although animal thyroid was given empirically to thyroid-deficient patients by the Chinese thousands of years ago, the story of the modern understanding of this gland begins when it was found that iodine has a beneficial effect in cases of thyroid-deficiency disease.

The thyroid produces an iodine-containing hormone called thyroxine which regulates the rate at which the metabolic processes in the body proceed. When the thyroid is sluggish or virtually extinct in adult life, the victim is cold, torpid, and loses hair, becoming anaemic and coarse-featured. This condition is called myxoedema because of the doughy swelling of the skin. Thyroid failure in infancy causes cretinism. When too much hormone is produced, the victim becomes nervous, excitable, hot, flushed, and in danger of heart failure from literally "burning himself up." This is the classical Graves' disease.

Robert Graves was one of the leaders of a brilliant group of Irish doctors in Dublin who reformed clinical teaching in Ireland and made many outstanding contributions to medical science. In 1835, Graves published his account of exophthalmic goitre. His description was so outstanding that Armand Trousseau, a French physician and a great admirer of Graves, proposed that the condition should be named Graves' disease—a lasting clinical eponym.

In the middle of the nineteenth century, it was shown that in animals thyroid deficiency could be produced by surgically removing the gland and, more important, that the condition could be reversed by regrafting thyroid tissue or by giving thyroid extract.

The connection of iodine with the thyroid gland was not understood until 1889 when the French physiologist Eugene Gley demonstrated the presence of iodine in the gland.

Beginning in 1912, Edward Kendall, a biochemist at the Mayo Foundation, studied the active principle of the thyroid. Two years later he obtained a substance which he called thyroxin and established it as an iodine-bearing hormone.

The parathyroids, usually four in number, are tiny glands under the thyroid. They secrete parathormone (PTH) which is involved in regulating calcium levels in the blood and, indirectly, in controlling blood phosphate levels. Low calcium levels can lead to muscle cramps and tingling in the limbs, while high calcium levels can cause muscle weakness.

The glands were first seen in 1862 by the English anatomist Sir Richard Owen, during a post mortem on a hippopotamus at London Zoo. But he did not get the credit for this. The parathyroids are usually said to have been discovered in 1879 by Ivar Sandstrom, a student at Uppsala.

In 1891, Eugene Gley described the parathyroid glands, demonstrated that the removal of the minute structures is fatal, and emphasized the need to preserve them during surgical removal of the thyroid for Graves' disease. Soon afterwards, it was discovered that the glands are connected with tetany, the spasm of the muscles due to calcium deficiency, and it was shown that extracts of parathyroid could relieve these spasms. It was not until 1926, however, that the substance parathormone was isolated from the glands.

The adrenal glands lie above the kidneys and are in two distinct parts—the outer cortex and the inner medulla. The yellow outer cortex produces glucocorticoids which affect protein and carbohydrate metabolism, mineralocorticoids, which regulate the concentration of fluid in the body, and the sex hormones—androgens and estrogens. The brown centre, the medulla, is totally different, both structurally and embryologically. It is closely linked with the autonomic nervous system and produces two hormones—adrenaline and noradrenaline—which are released in situations that engender such strong emotions as fright, fear, and anger, to raise the blood pressure, dilate the pupils, and to make the body instantly ready for "fight or flight."

The adrenal glands seem to have been unknown until 1563, when they were described by Eustachius. It is probable, however, that they had been noticed much earlier without being examined or referred to and certainly without being understood.

Eustachius, who called them *glandulae renibus incumbentes*, described the general shape and situation of the glands, but offered no explanation of their function. The first real contribution to the knowledge of the function of the adrenals was made by Thomas Addison in 1855, who at the same time made a contribution to the understanding of the whole endocrine system.

Harvey Cushing, who advanced neurosurgery to modern standards, lived from 1869 to 1939. Fourth in a line of physicians, he graduated from Harvard and studied abroad. As professor of surgery at Johns Hopkins, Harvard, and Yale, he combined pathology and a study of clinical diseases with surgical technique. A brilliant neurological surgeon, Cushing was also a prolific and gifted writer. The papers he presented were filled with reflections on current medical problems. His biography of Sir William Osler, the great clinician, was published in 1925 and won the Pulitzer Prize for literature.

Robert Graves, founder of the Park Street Medical School in Dublin, was born in 1796. After seeing the innovations in medical studies in Europe, he introduced new clinical practices to his students. Graves put scientific observation at the centre of his reform. The first to describe completely the disease that bears his name, he also used a watch to time the pulse and devised a system of clinical clerking. Rejecting the theory that fever is a disease to be treated by starvation, he recognized it as a symptom of infection. Before his death in 1853, Graves asked that his epitaph should read, "He fed fevers."

In that year of 1855, Thomas Addison, a physician at Guy's Hospital in London, described patients in whom tuberculosis had destroyed the adrenal glands in a monograph entitled "On the Constitutional and Local Effects of Disease of the Suprarenal Capsules." The importance of his discovery was not realized; it was actually the inauguration of the study of internal secretion and the ductless glands. The condition described by Addison was later named Addison's disease by the French physician Armand Trousseau, and it has remained one of the great clinical eponyms.

Charles Brown-Séquard, a British physiologist and neurologist, followed up some of Addison's work. By surgically removing the adrenal glands of animals, he produced the disease identical to that described by Addison. Brown-Séquard showed that the glands are essential to life, but he mistook their true func-

Thomas Addison, the London physician who initiated the modern science of endocrinology, was born in 1793. His early research on toxicology, published in 1829, was the first book in English on the subject. He also pioneered the use of static electricity to treat palsy, epilepsy, and spasticity. In 1837, Addison was appointed physician at Guy's Hospital where he became joint lecturer in medicine and, with his colleague Richard Bright, wrote *The Elements of the Practice of Medicine.* Addison studied skin diseases, pneumonia, and tuberculosis. In 1849, he described pernicious anaemia, which was often known as Addisonian anaemia. In 1855, he published an accurate clinical description of the features of adrenal insufficiency, which later became known as Addison's disease. Despite his reputation as a brilliant lecturer and diagnostician, Addison was not popular with his colleagues and

no one in Britain took notice of his important work. Criticism and even derision were levelled at him, together with personal slights against his professional integrity. In 1860, Thomas Addison committed suicide.

Charles Edward Brown-Séquard was born in Mauritius, educated in France, and, although he practised and lectured in France, England, and the United States, he always kept his British citizenship: In London he was physician to the National Hospital for the Paralysed and Epileptic. In America he was professor of physiology and of pathology of the nervous system at Harvard. In 1872 he returned to Paris and succeeded Claude Bernard as professor of physiology. Brown-Séquard, who published more than five hundred papers, did work on the nervous system as well as on internal secretions. He died in Paris in 1894.

tion. He thought, as did many others after him, that the adrenals removed certain poisons from the blood, rather than adding substances to it. It was not until 1913 that it was proved that it is the cortex of the adrenal, not its medulla, that is indispensable.

The first extract of adrenaline, the panic substance, was made in 1894 by George Oliver and Sir Edward Sharpey Schaefer, the brilliant English physiologists who had been the first to isolate the active principle of a ductless gland and to investigate its physiological effects.

Edward Kendall, of the Mayo Foundation, did important work on the identification of the hormones of the adrenal cortex and isolated the so-called Kendall's Compound E, which in 1949 became known as cortisone. It was demonstrated not long afterwards that cortisone is effective in the treatment of many illnesses, including rheumatoid arthritis, and the substance gave promise of becoming a miracle drug, until the dangers of continual dosage were recognized.

The pancreas, the large gland in the abdomen, produces insulin, which decreases levels of sugar in the blood, and glucagen, which increases blood sugar levels. The control of fuel in the body is a balance of these two hormones, together with other hormones, providing a regulated supply of fuel to all cells for metabolism and growth in all conditions. It is the inadequate production of the hormone insulin which indirectly results in diabetes mellitus. It was an inspired guess on the part of a London physician, named Cawley, that led to the publication in 1788 in the *London Medical Journal* of the theory that the pancreas is involved in diabetes mellitus.

Diabetes is a word derived from two Greek words, one which means a siphon and the other to pass through. Aretaeus of Cappadocia, a Greek physician who lived in the second century A.D., gave the condition this name because of the excessive urination which is a symptom of it. He wrote: "The epithet diabetes has been assigned to the disorder being something like passing of water by a siphon."

The term was used by Galen, but he said he had only seen the condition twice. It was the excessive urination, polyuria, that attracted the attention of the early observers. It was recognized that some people had polyuria with urine that contained sugar, while others had polyuria with urine that was clear and insipid.

Thomas Willis, the seventeenth-century English physician, first rediscovered the sweet taste of urine as a test for diabetes—a test which had been known to the Chinese a thousand years before. It was Willis, too, who differentiated diabetes mellitus, meaning honey sweet, from diabetes insipidus, but he did not know what caused either condition.

It was the pancreas that made Claude Bernard wonder if some internal secretion played a part in the regulation of sugar metabolism. Between 1848 and 1857, this extraordinary physiologist at the College de France in Paris isolated glycogen in the liver and put forward his theory of the *milieu intérieur.*

In 1899, Joseph von Mering and Oskar Minkowski, two famous men in the annals of endocrinology, proved experimentally the connection of diabetes with the pancreas by observing dogs whose pancreas had been removed. As the result of their experiments, there remained no question that some substance secreted by the pancreas keeps diabetes at bay.

The actual sites of production of insulin were first seen in 1869 by Paul Langerhans, the German physician, anatomist, and pathologist, but this was a histological observation and he had no idea of the function of the million clumps of cells buried in the exocrine part of the pancreas that now bear his name as the islets of Langerhans. It was almost twenty-five years later that the connection was made between the islet cells and the formation of this antidiabetic substance.

Not until the 1920s did anyone succeed in isolating the hormone insulin and on this hangs a mysterious tale. Nicolas Paulescu, the Romanian professor of physiology, extracted a hormone which he called pancreine and proved that it relieved diabetes in dogs whose pancreas had been removed. Paulescu worked out the actions, the dosage, and the side effects on protein metabolism of this hormone and, in 1921, described his findings in a lecture and in a paper published in a widely known French journal of physiology. Almost a year later, however, Frederick Banting and Charles Best, Canadian physiologists at the University of Toronto, with knowledge of Paulescu's publication, published virtually the same work. They called their extract insulin and, with the help of the huge American pharmaceutical company Eli Lilly, produced insulin on a commercial scale. The 1933 Nobel Prize in medicine for the discovery of insulin was awarded not to Paulescu nor to Best, however, but to Banting and to J.J.R. McLeod, who was the director of the Toronto laboratories.

The whole story is riddled with ambiguities and suspicion. Poor Paulescu died some years later, of a broken heart, it is said, because of the lack of recognition for his work. The discovery of the pancreatic hormone ranks with antibiotics as one of the great life-saving discoveries of all time.

The male and female gonads, or sexual organs, are among the body's major endocrine glands. The testes produce testosterone, which maintains the male secondary sexual characteristics. The ovaries produce progesterone, which helps to prepare the uterus for pregnancy, and estrogen, which produces the female secondary sexual characteristics.

In 1660, Thomas Willis, whose medical interests were so varied, was perhaps the first man to suggest that masculinity is a product of the testes and, without being able to prove it, speculated that they secrete some "masculinity factor" into the blood.

Although the sexual organs were among the first to be suspected of having an endocrine function, modern work did not begin until 1889, when Brown-Séquard experimented with testi-

Edward Sharpey Schaefer was one of the group of outstanding physiologists in Britain who pioneered in endocrinology. He was born in 1850, and at the age of twenty-one he became the first recipient of the Sharpey Scholarship at University College, London, where he studied with the great anatomist and later did research and taught. He held the chair of physiology at Edinburgh from 1899 until 1933, two years before his death. The inventor of a method of resuscitation known as Schaefer's method, he received many honours, including a knighthood. Schaefer added his mentor's name as a prefix to his own in 1918.

cular grafts in roosters. He found that the failure of a cock's comb to grow in a castrated fowl could be reversed by implanting grafted organs. And it was not until 1927 that a male hormone was first extracted. The female hormone estrogen was isolated three years earlier from the fluid from the ovarian follicle.

Other structures have functions suspected, rather than proven, to be endocrine. Among these has been the thymus which is large in the infant, but shrinks at puberty. Although its role in immunity, as part of the reticulo-endothelial system, is understood, the hormonal function of the thymus is not clear. Recently, however, the hormone thymic humoral factor (THF) was extracted. It is now believed by researchers that this hormone can be used to treat virus infections and cancer, and might even be able to slow down the aging process. If THF lives up to their expectations, it could become one of the most important drugs of the 1980s.

The pineal gland in the brain is still something of an enigma. That this structure has endocrine functions is obvious, but its action is less clear than that of other glands and no definite hormones have yet been isolated or synthesized. Nevertheless, even the earliest discoveries of the anatomy and physiology of the endocrine system have led to exciting and important developments both in methods of treatment of hormone-related conditions and in the understanding of the hormonal control of Claude Bernard's *milieu intérieur*. Although the endocrine function of the pineal body is not yet understood, we have come a long way since the pineal was believed to be the seat of the soul or since Descartes's view of it as the focal point for the operation of the nervous system was accepted.

Male Generative Organs

Annotated in Latin, this diagrammatic scheme of the male generative organs appeared in an English medical manuscript of the late thirteenth century.

The gross anatomy of the male reproductive apparatus has been well known since antiquity. This is to be expected, because of the visual prominence of these appendages and also, perhaps, because of man's age-old preoccupation with sexual matters, particularly with outward evidence of virility.

Aristotle's famous lost diagram, the first anatomical illustration, included, according to his description of it, an accurate depiction of the internal male organs. In addition to the urinary system, the diagram showed the two testes and their connections with the urethra. Writing in Greek, Aristotle called the testicle by the name *orchis*, which was also used for an olive and for the root of the plant, the orchid, which has roots shaped like a testicle. Today the derivation is still in medical use. Inflammation of a testicle, for example, is called orchitis. The word testis is of Latin origin and means to bear witness, in this case to the maleness of the owner.

Aristotle's drawing showed the testicular

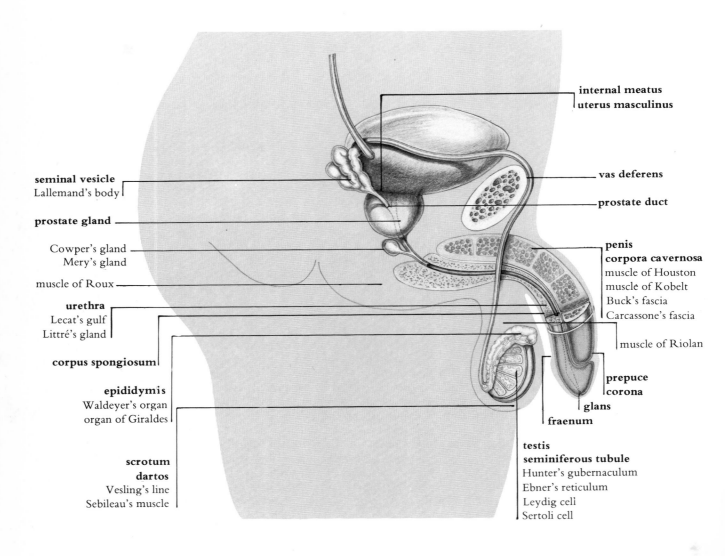

seminal vesicle
Lallemand's body

prostate gland

Cowper's gland
Mery's gland

muscle of Roux

urethra
Lecat's gulf
Littré's gland

corpus spongiosum

epididymis
Waldeyer's organ
organ of Giraldes

scrotum
dartos
Vesling's line
Sebileau's muscle

internal meatus
uterus masculinus

vas deferens

prostate duct

penis
corpora cavernosa
muscle of Houston
muscle of Kobelt
Buck's fascia
Carcassone's fascia

muscle of Riolan

prepuce
corona

glans

fraenum

testis
seminiferous tubule
Hunter's gubernaculum
Ebner's reticulum
Leydig cell
Sertoli cell

artery and vein, the two long blood vessels which supply the testes. His anatomy was correct in detail, for the right testicular vein drains into the inferior vena cava, the main central vein trunk of the abdomen, and he correctly noticed that the left vein goes up to the kidney region and joins the left renal vein—a remarkable observation considering that it was made in the fourth century B.C.

Aristotle next drew the epididymis, which he called the "sinewy duct extending along the orchis." The epididymis is the long, contorted tube which carries the sperm away from the site of production in the testis. It wriggles its way from the top of the testis down to the bottom, before straightening out to sweep back upwards as the vessel which Aristotle called the "returning duct"; it conveys the spermatic fluid to the neck of the bladder for emission through the urethra. This spermatic duct is today known as the vas deferens, the term introduced in the early sixteenth century by the Italian anatomist Jacob Berengar.

Around the testis and epidymis is a double bag called the tunica vaginalis. It lies inside the bag of skin which is the scrotum, meaning sack. Aristotle called it the "membrane enveloping the orchis and returning duct."

Finally, the Greek philosopher-scientist described the inner part of the urethra, the tube which leads both urine and semen to the outside, and the penis, which means tail and was one of the many slang terms used by the Romans for the male organ. Among the words the Romans used was *clava*, meaning club, *gladius*, the word for sword, and *vomer*, meaning plough. The word penis, derived from the Latin *pendere*, to hang down, was most commonly used. It was, therefore, adopted as an anatomical term and has been used as such in English since the seventeenth century.

After Aristotle there was little left to describe in the appearance to the naked eye of the male genital organs. Certainly, the poor draughtmanship and worse observational powers of the early medieval anatomists contributed nothing new. The revival of descriptive anatomy in the fifteenth century, however, brought further clarification of the male reproductive paraphernalia, as it did of other body systems.

Leonardo da Vinci drew the uro-genital systems of both sexes in schematic form and in beautiful perspective, although he made a number of factual errors, such as the incorrect origin of some blood vessels. The muscle men in the woodcuts and copperplate engravings of later anatomists all had prominent and accurately drawn external genitals, although for some time there continued to be mistakes in the depiction of the internal structures.

There were excellent illustrations in Andreas Vesalius's *De Fabrica*. In these the portrayal of

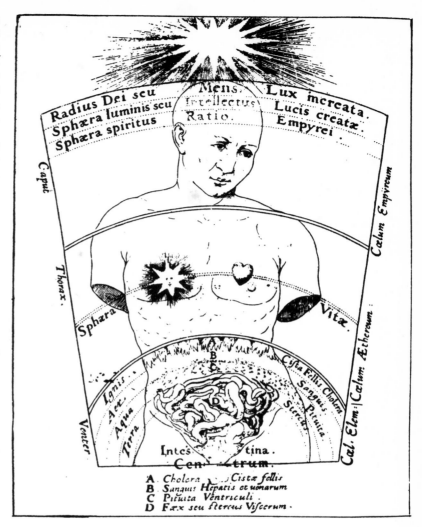

A. *Cholera* : *Cistæ fellis*
B. *Sanguis Hepatis et uenarum*.
C. *Pituita Ventriculi*.
D. *Fæx seu Stercus Viscerum*.

the gonads—the alternative name for the generative organs of both sexes—of the male were entirely accurate.

In the male the region around the neck of the bladder is anatomically rather complex. Here the two ureters enter from the kidneys and the internal opening of the urethra forms the third point of the trigone of the bladder. Around the neck of the bladder the walnut-sized prostate gland wraps its three lobes. The prostate was named appropriately by Erasistatus in the third century B.C. The word comes from the Greek for "one who stands before," which is exactly what this gland does in relation to the exit from the bladder.

Above and behind the prostate, the two seminal vesicles are the site of production of most of the volume of seminal fluid when sexual ejaculation takes place. The sperm themselves, stored in the epididymis alongside the testis, constitute only a tiny part of the ejaculate. When ejaculation occurs, the seminal vesicles eject a fluid rich in fructose, a sugar which supplies the energy for the sperm to undertake their long

Robert Fludd, who lived from 1574 to 1637, was an English alchemist and a mystic. His vision of man—the complete being who would be the model for all humanity—was an extremely intricate one, in which various anatomical and spiritual features were curiously interwoven. The illustration above of the Cabalistic Adam as the first man and archetype of humanity, appeared in Fludd's *Collectio Operum*.

Discovering the Human Body

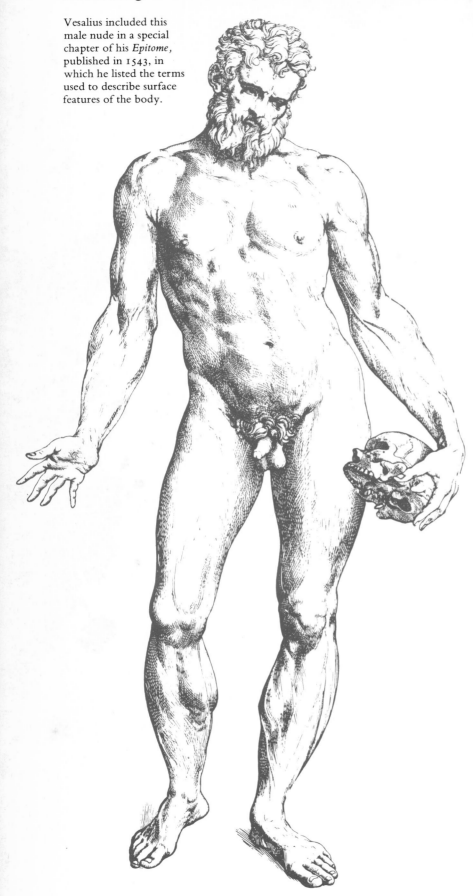

Vesalius included this male nude in a special chapter of his *Epitome*, published in 1543, in which he listed the terms used to describe surface features of the body.

journey up into the uterus, where one of the cells may find an ovum.

The spermatozoa are motionless while stored and it is the admixture with seminal vesicle fluid and such other glandular secretions as that from the prostate, that initiates their ceaseless flagellant motion.

Mystery still surrounds the curious granules found in the seminal vesicles. Lallemand's bodies are tiny amorphous concretions, rather like sand. They were described in 1825 by Claud François Lallemand, a French surgeon with a particular interest in genito-urinary diseases. The function of Lallemand's bodies remains obscure, although one theory is that they could be the result of precipitation of some of the seminal fluid which is manufactured and stored in these vesicles.

The two seminal vesicles send their ducts into the first part of the urethra, which is buried in the prostate. The vas deferens on each side and the openings of the prostatic ducts also empty into this section. When ejaculation is about to occur, the entry into the bladder above is shut off by muscular contraction. The actual orifice of the urethra into the bladder is called the internal meatus.

A tiny cul-de-sac in the back wall of this part of the urethra is called the uterus masculinus. As its name suggests, it is the analogue of the female womb, although there is some learned argument as to whether it more accurately represents the vagina. In the same way as the male has his womb, or vagina, so the female has the homologue of the penis in her clitoris. Because most of the sexual organs have embryological counterparts of the opposite gender, in some malformations of foetal development confusing degrees of anatomical bisexuality may occur.

In the urethra, there are a pair of glands, opening into the upper portion, which are known as Cowper's glands, although an alternative eponym with a claim to priority is Mery's glands. William Cowper, a London surgeon, who devoted much of his time to anatomy, described the bulbo-urethral glands in 1702, fifteen years after Jean Mery, chief surgeon to the two main hospitals of Paris and to the Queen of France, had described them before the French Academy of Medicine.

Cowper acquired a reputation for purloining the anatomical work of others without reference to the original. As well as successfully advancing his claim over Mery, Cowper heavily plagiarized the work of Govert Bidloo, who published a magnificent atlas of anatomy. As Jessie Dobson, the curator of the Museum of the Royal College of Surgeons, rather delicately says, "Cowper was much indebted to the work of Bidloo which he failed to acknowledge." Nevertheless, Cowper, a rich and successful surgeon, became a fellow of the prestigious Royal Society in 1698.

A curious name for a dilated part of the lower urethra is Lecat's gulf. It commemorates Claude Nicholas Le Cat, a surgeon and anatomist in Rouen in the eighteenth century.

The French are well-represented in the male generative organs. The urethral glands of Littré are named for the French surgeon and anatomist Alexis Littré, who, in 1700, described these small glands in the lining of the urethra.

A structure with two competing eponyms is the paradidymis, a small remnant of embryological tissue. It lies against the testis and is known either as the organ of Giraldes or Waldeyer's organ. Joachim Albin Cardazo Cazado Giraldes has priority, for he wrote on this organ in 1859. Born in Porto, Portugal in about 1808, Giraldes qualified in medicine in Paris and spent the rest of his life in France. A professor of surgery, he died in 1875 of septicaemia, the result of cutting himself while performing a post-mortem examination. Heinrich Waldeyer, a professor of pathological anatomy in Berlin and Breslau, has a number of other eponymous claims—in the brain, the ovary, and in the throat—but his description of the paradidymis came long after Giraldes.

The phallic organ itself has a complex structure, with two main distensible tubes, called the corpora cavernosa, surmounting a compressible tube known as the corpus spongiosa. The veins, which drain the cavernosa, can be obstructed by reflex nervous and muscular mechanisms. This cuts off, or at least impedes, the return of blood drainage from the corpora. But since the higher arterial pressure is still forcing blood in, the result is stiffening, or erection, necessary for sexual activity. If the urethra ran unprotected alongside the corporae cavernosae, the pressure within the skin covering would also obliterate its cavity and defeat the object of the exercise because the seminal fluid could not be ejected. The corpus spongiosa, therefore, surrounds the urethra and acts as a buffer, for its compressibility permits the urethra to remain open.

The end of the penis is called the glans, meaning acorn, the same name that is used for the head of the clitoris in the female. The fold of skin which is removed at circumcision is the prepuce and this is anchored underneath the shaft by a narrow band of skin called the fraenum, meaning a bridle. The projecting edge of the glans is the corona, or crown.

The deep fascial sheath of the penis is known as Buck's fascia, commemorating Gordon Buck, a New York surgeon, who described it in detail in 1848. Although Buck's name is remembered in this fascia, his greatest gift to medicine was the development of extension and traction for the treatment of fractures of the thigh.

Another fascia deep at the base of the penis is known as Carcassone's fascia and is named for Maurice Carcassone, a Frenchman, who was

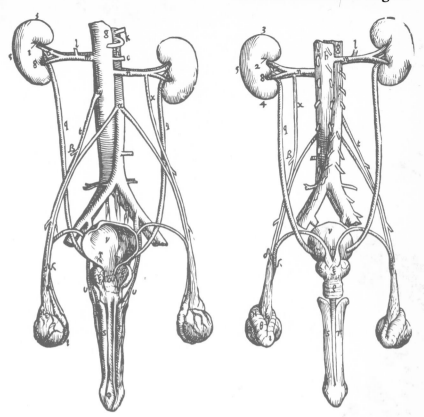

professor of anatomy at Montpellier in the nineteenth century.

A number of muscles related to the penis are named. They include the muscles of Riolan, of Roux, of Houston, and of Kobelt, but modern nomenclature has pretty well retired these along with many other eponyms.

Down the midline of the scrotum is an obviously thickened line, called the raphe, or seam, which is exactly what it is. This is sometimes called Vesling's line, after Johann Vesling, a German who was appointed to Vesalius's old chair in 1632, and who also served as the director of the almost equally famous botanical garden in Padua.

Under the skin of the scrotum is a thin sheet of muscle called the dartos. The two testicles share the scrotum, but a central partition separates them and part of this is muscular and is known as the intertesticular dartos, or Sebileau's muscle, because Pierre Sebileau, a Parisian anatomist, described it in 1897.

All the scrotal muscles can contract and tighten on the testes during sexual activity, presumably to help eject the seminal fluid by adding pressure from without. In the unborn male child, and indeed in the infant, the testes are not in the scrotum, but are high in the abdomen, almost in the same position in which the ovaries remain throughout a woman's lifetime. As maturation proceeds, the testes are drawn down into the scrotum.

It seems that spermatogenesis, the develop-

Two plates from the fifth book of *De Fabrica* give accurate front and rear views of the male urogenital apparatus. In both, Vesalius showed the kidneys to either side of the major blood vessels of the lower trunk, the descending aorta, and the inferior vena cava. In the front view, left, the bladder and the urethra, the common passage which conducts urine and semen to the outside, have been laid open.

The seventeenth-century Dutch anatomist Regnier de Graaf made an extensive study of the male organs. In a treatise, published in 1668, on the male reproductive system, in which this illustration appeared, he described the prostate gland and the seminal vesicles.

William Cowper, immortalized in the bulbo-urethral glands, was born in 1666. A London surgeon, he devoted most of his energy to anatomizing and in 1698 published his *Anatomy of Human Bodies* "in sumptuous fashion." The text was Cowper's own, but most of the plates were not. Bidloo, whose illustrations had been borrowed without permission—or credit—promptly excoriated Cowper in an open letter to the Royal Society. In 1702, seven years before his death, Cowper wrote an apology entitled "Eucharistia," which included a description of the glands. It was later learned that they had already been described by Mery before the French Academy.

ment of sperm, can only take place at a slightly lower temperature than that inside the body cavity. This is the reason the testicles are hung outside the body in a cooler, but much more vulnerable and inconvenient, situation.

The process which draws or guides the immature testis downwards is Hunter's gubernaculum, a fibrous band which appears to physically drag the organ downwards as growth proceeds. The word gubernaculum is derived from the Latin word for governor, as it presumably governs the descent. Hunter is John Hunter, the great Scottish anatomist and surgeon who described it in 1762.

Although the basics of male anatomy were there to be read from long ago, it was in the understanding of the whole process of reproduction that hypotheses and debate flourished for millennia.

Hippocrates, a century and more before Aristotle, had some views on generation—what today is known as embryology—although it is more likely that it was his son-in-law Polybus who originated the ideas. The concept was that there were two seeds, one, the semen, from the male and the other, the vaginal secretion, from the female. This was in conflict with the later theory of Aristotle that a new life came only from the male. He believed that semen, although mixed with the woman's menstrual blood, was the sole contributor to the formation of the embryo. This concept relegated the physiological role of the woman to one of merely nourishing the new life and began a controversy which was to stunt the growth of scientific embryology for many centuries.

Inextricably linked with the study of embryology was clarification of the microstructure of the testis, the fundamental male organ of generation. The most important discovery emerged with the development of the microscope. It was that pioneer of optics Anton van Leeuwenhoek who first identified and described the spermatozoon, the male sperm, an achievement which was to make him famous above all his other extensive work.

In March 1678, Leeuwenhoek first described what he called spermatic animalcules in a letter, with accompanying drawings, to Nehemiah Grew, the secretary of the Royal Society in London. A year later, this material was published in the *Philosophical Transactions*, perhaps the best-known scientific journal in Europe at that time. It is amazing that Leeuwenhoek could have made such exact drawings of the sperm magnified more than two thousand times, when his microscope was indeed a crude instrument with only a single lens.

The next significant contribution was made by Lazzaro Spallanzani, the Italian priest and physiologist, who, in 1786, was able to show that the spermatozoon is essential to the fertilization of the female egg. This fundamental discovery helped to dispel the centuries-old concept of spontaneous generation.

Spermatozoa were recognized as cells by the nineteenth-century Swiss anatomist Rudolf Albert von Kölliker, who located the formative site of the sperm in the testis and investigated the development of spermatozoa.

Sperm are produced in the numerous seminiferous tubules which lie in groups within the testis. The words semen and seminiferous derive from the Latin to sow or pertaining to seed. The word sperm has exactly the same meaning, although it comes from the Greek.

A network in the seminiferous tubules, sometimes referred to as Ebner's reticulum, was described in the latter part of the nineteenth century by an Italian, Victor Ritter von Rosenstein Ebner, who was born in Breganz, a German-speaking area of Italy, and became a professor of anatomy at Innsbruck in Austria.

Enrico Sertoli, an Italian physiologist and professor of experimental physiology at the University of Milan, carried out fundamental research on the way in which sperm are formed and on their mobility, a vital factor in fertilization. He described the formation of sperm and is eponymized in the Sertoli cells, which are the cells which support the cells in the testicular epithelium that produce spermatozoa. His account of his work was published in 1865.

Other cells found between the sperm-producing tubules are known as Leydig cells, commemorating the German Franz von Leydig, one of the great nineteenth-century pioneers of comparative histology. In 1850, he described these cells which he recognized as the site of production of the male hormone testosterone, although the actual endocrinological details were not understood until much later.

John Hunter was born in Scotland in 1728. He never studied medicine, but at the age of twenty he went to London to work with his elder brother William in his school of anatomy, thus beginning a career as an anatomist and surgeon. In 1768, he became surgeon to St. George's Hospital, a post he held until his death in 1793. While doing his own research, Hunter innoculated himself with syphilis and this undoubtedly caused much of the illness he suffered in later years. After his death his collection of anatomical and pathological specimens formed the basis of the museum of the Royal College of Surgeons.

Rudolf Albert von Kölliker, who lived from 1817 until 1905, spent his whole life in Switzerland. For fifty years he was professor of physiology and comparative anatomy at Würzburg where he was influential in developing the university as an important medical centre. Von Kölliker showed that smooth muscle is composed of nucleated muscle cells, described the branching of cardiac muscle, the grey matter around the spinal canal, and the iris of the eye. He expanded Corti's work on the ear and investigated the development of spermatozoa. In addition, he edited a journal of zoology for fifty years and wrote a textbook on microscopic anatomy.

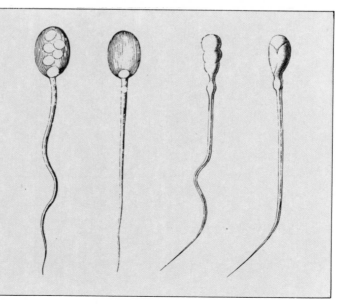

Anton van Leeuwenhoek was the first person to observe many microscopic organisms by means of a lens. Although his work lacked the discipline and organization of formal scientific research, his untiring observations led to discoveries of enormous importance. A young student named Johann Ham worked with him and together they discovered spermatozoa. In 1677, Leeuwenhoek communicated their findings to the Royal Society in London with his own drawings of the spermatozoa in man, right, and in dogs, far right.

Female Generative Organs

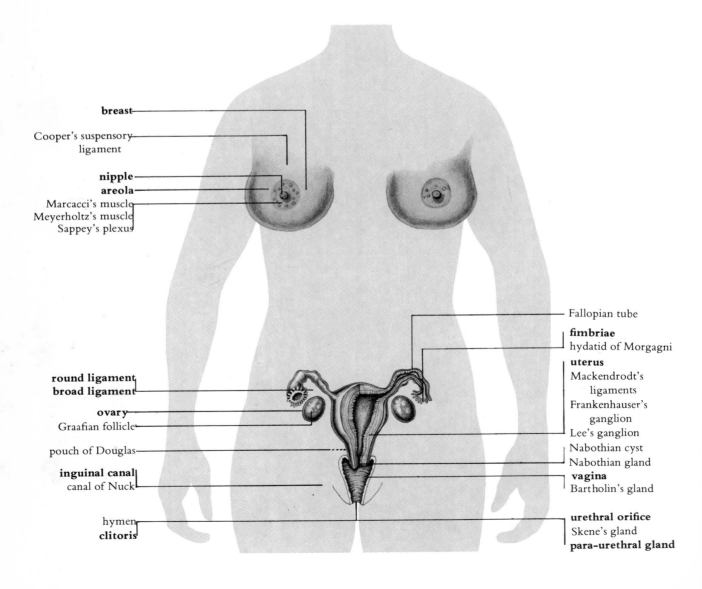

Soranus's diagram of the uterus, although bicornate, was amazingly accurate, considering that he practised in about A.D. 100.

Knowledge of female anatomy and physiology lagged far behind that of the male, until the sixteenth century when Leonardo da Vinci produced his beautiful and detailed studies of the female form. Until the Renaissance, propriety dictated that a woman's body could not be represented unclothed even in the interests of science. The exception was the pregnant woman. Pregnancy, after all, was evidence of the importance of a woman to society, as the carrier of the next generation, and of her responsibility as childbearer. Therefore there was justification for the study of the female body, if only to gain understanding of the reproductive system.

The most important structure in female reproductive anatomy, the uterus, was represented in an Egyptian hieroglyph as early as 2900 B.C. Even in this conventionalized form it was anatomically more accurate than representations which were made more than four thousand years later. In early Greek medicine

breast
Cooper's suspensory ligament
nipple
areola
Marcacci's muscle
Meyerholtz's muscle
Sappey's plexus

round ligament
broad ligament
ovary
Graafian follicle
pouch of Douglas
inguinal canal
canal of Nuck
hymen
clitoris

Fallopian tube
fimbriae
hydatid of Morgagni
uterus
Mackendrodt's ligaments
Frankenhauser's ganglion
Lee's ganglion
Nabothian cyst
Nabothian gland
vagina
Bartholin's gland
urethral orifice
Skene's gland
para-urethral gland

the uterus, which is tiny and pear-shaped in repose, was misrepresented and much maligned. Early portrayals show the human womb as bicornate, two-horned, as it is in other mammals. This error persisted until the sixteenth century.

Four hundred years before the birth of Christ, no less a figure than Plato instigated the myth of the wandering womb. He believed that, unused for a long time, the womb became "indignant" and made its way around the body, inhibiting the body's "spirits" and causing disease. This confused concept influenced the way men thought about women—that they were irrational, for example, and prone to emotional outbursts. The very word hysteria comes from the Greek *ustera,* meaning uterus. The early Greeks were convinced—and the conviction continued for centuries—that women were inevitably unstable because of the anatomy peculiar to their sex.

Although Aristotle had described the uterus, it was more than four hundred years later that it was depicted with reasonable accuracy by Soranus of Ephesus, an outstanding physician who lived from A.D. 78 to 120, studied at Alexandria, and practised in Rome, apparently specializing in obstetrics and gynaecology. He wrote a sophisticated treatise entitled *On Diseases of Women* in which the account of the uterus is regarded as an outstanding piece of ancient descriptive anatomy, despite his description of the shape of the organ as bicornate. Soranus also made an intensive study of prolapse of the uterus and claimed that the uterus neither wandered nor caused disease.

Unfortunately, Soranus, who included in his treatise a diagram showing the anatomy of the pregnant woman, had no direct influence on the course of anatomy, for his text was lost until the nineteenth century. Consequently, the ancient Romans relied on speculation about the likely structure of the womb. One of their greatest anatomists claimed that it was an organ consisting of no less than seven chambers. It was not until seventeenth-century anatomists began systematic dissections of the human body that the size and shape, and eventually the mode of function, of the uterus became known.

The ancients, of course, acknowledged and named what could be seen and easily understood. Thus ancient Roman ribaldry yielded the anatomical term vagina, the Latin word for the sheath, or scabbard, of a sword. The early Greeks perceived the role of the clitoris as the focus of woman's sexual arousal. From a language in which one word so often encapsulates a whole concept came *kleitoris*—from *kleis,* meaning key.

So many parts of the body, including the female organs, carry the names of men, it is perhaps fitting that here a woman is com-memorated. But even this woman is mythical, for it is Venus, the goddess of Love. This female eponym is Latinized to name the mound of fibrous and fatty tissue that clothes the pubis. The mons veneris, the mount of Venus, is the hair-covered pad that protects and conceals the deeper cleft. The origins of the name are lost, but it was probably first used during the great anatomical resurgence of the Renaissance.

The name of the delicate membrane which partly closes the lower end of the vagina and demarcates the virgin state derives from Greek mythology. Like the mons veneris, the hymen commemorates a deity, although this time a god, rather than a goddess. Hymen was the god of marriage and nuptial harmony. There is an Attic legend which depicts him as a beautiful youth who follows a maiden to Eleusis; disguised as a woman, he saves her from pirates and wins her as his bride. Perhaps it is the tenor of other stories of Hymen—as an effeminate youth, sometimes carrying a bridal torch and wearing a veil—that more appropriately links his name to this ephemeral structure which conventional morality dictated should only be penetrated on the wedding night.

Just as Plato was so terribly wrong about the womb, so Aristotle, the great philosopher, helped perpetuate another myth that was to haunt the literature of reproductive anatomy for centuries. This was the belief in the existence of female semen. Like Aristotle, Herophilus of Alexandria mistook the copious lubricant exuded during sexual arousal for female seminal fluid. Although all his writings were lost, he is quoted by Galen as maintaining that the fluid

Egyptian hieroglyphs representing the uterus are often seen on representations of Tuart, a hippopotamus-headed goddess associated with childbirth.

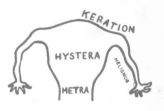

This depiction of the uterus, according to Aristotle, was fanciful. His nomenclature has been retained, as has the emotional bias of the associations, and women are still considered prone to hysteria.

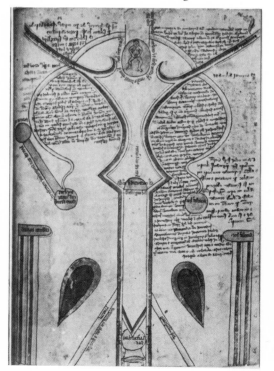

This schematic drawing of the womb, from a manuscript of the late thirteenth century, shows that concepts of fertilization and generation were vague. At the top centre of the drawing is the usual homunculus, the little man, awaiting quickening and release.

arose in the woman's testes, by which he meant the ovaries. According to Herophilus, this semen then travelled by tiny ducts to the neck of the bladder.

Implicated in this confusion were several groups of small glands in the region of the outlet of the urethra and of the vaginal orifice which were long thought to be the source of the lubricant fluid. The para-urethral glands lie on each side near the lower end of the urethra. Numbers of these tiny glands are grouped together and open into ducts—the para-ure-thral ducts. Each duct runs down into the sub-mucous tissue and ends in a small opening on the edge of the urethral orifice. The first to describe these small glands was the brilliant young Dutchman Regnier de Graaf. His note on them, however, was an almost incidental part of his much larger and beautifully accurate account of the anatomy of the female pelvic area, *De Mulierum Organis,* published in 1672. In this he mentioned the para-urethral ducts as a focus of the venereal disease gonorrhea.

Only five years later, the Danish anatomist Caspar Bartholin published his account. The son and grandson of famous medical pioneers, Bartholin gave full credit to de Graaf's earlier description of the para-urethral ducts when he published his *De Ovariis Mulierum* in 1677. But he went on to discuss the two small glands embedded in the minor labia at the entrance to the vagina. Bartholin claimed that it was through these, the vulvovaginal glands, that the coital fluid was emitted, rather than through the urethra, as de Graaf believed.

From then on, the vulvovaginal glands have been known as Bartholin's glands, and the name has passed so completely into medical usage that such terms as bartholinitis are well recognized in modern gynaecology. Indeed, the eponym has passed out of anatomical use into a more pathological context; Bartholin cysts and Bartholin abcesses are standard terms for the common infections that can affect these glands.

For almost three hundred years after Bartholin lent his name to the vulvovaginal glands they were presumed to contribute sufficient lubrication for successful and painless vaginal penetration. It was not until 1966, when Dr. William Masters and Mrs. Virginia Johnson published *Human Sexual Response,* the product of eleven years of research into the anatomy and physiology of human sexuality, that the scientific facts were finally known about these glands. Masters and Johnson were able to prove that the Bartholin glands produce only a minimum amount of material and that the lubricating fluid released immediately on arousal is exuded through the walls of the vagina.

Two hundred years after de Graaf and Bartholin had described the para-urethral ducts, they were discovered by a Scotsman, Alexander Johnston Chalmers Skene. He had emigrated to the United States and studied medicine at Long Island College Hospital, from which he was graduated in 1863. Skene became a professor of gynaecology there in 1870 and was subsequently elevated to dean and then to president. His eminence grew and he became professor of gynaecology at the New York Postgraduate Medical School, opened two private hospitals for women's diseases, and was a founding member of the American Gynaecological Society. But his name is known to doctors around the world for its attachment to the very glands which Bartholin and de Graaf had described all those years before.

In 1880, Skene published an article in the *American Journal of Obstetrics* announcing his discovery of the para-urethral ducts and stating that he had been "unable to find any reference to them nor to any pathological accounts of their diseases." He confessed that he had no idea as to their purpose or function. In any event, Skene's glands they became on the strength of one case which Alexander Skene had described —a woman of thirty who had an acute infection of the glands which Dr. Skene discovered, treated, and cured.

In about 1650, Guillaume des Noves, a professor of anatomy at Genoa, contributed to the endless argument about the source of female semen by describing small, fluid-filled blobs on the surface of the cervix, which he had observed from his dissections of three women. When he pricked these with a needle, a substance escaped "quite different from the oily humour that is secreted by women in the throes of amorousness," which he compared to male semen.

In 1707, however, Martin Naboth, a German physician and professor of chemistry, described mucous glands on the cervix in his book *De*

Caspar Bartholin was born in Copenhagen in 1655. His grandfather, Caspar Primus, was a physician and theologian, and his father, Thomas Primus, was a professor of mathematics who became an anatomist. Aided by his family's reputation, Caspar Secundus was appointed professor of philosophy at the University of Copenhagen at the age of nineteen. He did the customary European tour for three years and returned to become professor of physics in 1677, the year his anatomical treatise was published. An active and versatile man, he went on to gain many civic honours. He continued his research and teaching until his death in 1738.

In these illustrations from a thirteenth-century manuscript of medical case histories, a woman, apparently of high estate, is swooning and being revived. She is surrounded by female attendants, a physician, and a tonsured cleric. The feather held to the lady's nose in the bottom illustration has probably been dipped in an inhalent, an analogue of the later smelling salts. It is not unlikely that her swoon was attributed, as it would have been since antiquity and would be until much later, to the uniquely female and altogether mythical complaint of a wandering womb.

Sterilitate Mulierum. These glands frequently have small retention cysts, the blobs which des Noves described. Naboth thought, therefore, that they were egg sacs, or ova, and they were called ovula by Naboth in the mistaken belief that they were related to female fertility. Later, however, this was corrected and they were called Nabothian glands, while the cysts were known as Nabothian cysts or follicles.

It is quite probable that although their concept of the uterus itself was a confused one, the ancients had some idea about the uterine tubes. Galen wrote that Herophilus recognized them in the fourth century B.C. and, soon after the dawn of the Christian era, Rufus of Ephesus mentioned them in both sheep and humans. But he seemed to think that they were merely supporting ligaments, rather than vital con-

This illustration from a fourteenth-century manuscript shows the seven-chambered uterus—an excellent example of medieval anatomical fantasy.

ducting tubes for the release of the human egg.

It was left to Gabriele Fallopius, one of the greatest early Italian anatomists, to describe these egg tubes, or oviducts. In 1561, at the height of the anatomical renaissance, he published his *Observationes Anatomicae* in which he described the uterine tubes as "trumpets of the uterus." He recognized the function of these tubes as "seminal ducts" and likened them to "a brass trumpet, with bent parts like this classical instrument." At the ovarian end, the shape of the widening tube reminded Fallopius of the orifice of the trumpet. Fallopius's name soon became ineradicably linked with the uterine tubes and today it is not unusual to omit the capital letter in the fallopian tubes—a sure indication of the total absorption of an eponym into common usage.

The trumpet ends of the Fallopian tubes frequently have small cysts, or bladders, attached to the fine fimbriae, or fringes, that seem to reach out to envelop the nearby ovaries. These cysts are called hydatids, a word derived from the Greek for a drop of water, which quite aptly describes them. These hydatids had first been noted by Regnier de Graaf, but it was Giovanni Battista Morgagni, the Father of

There is some speculation that this female figure was drawn, or at least influenced, by Leonardo. It appeared in a late fifteenth-century Venetian book. Although the anatomy is largely inaccurate, an attempt was made to depict the uterus naturalistically.

Pathological Anatomy, who actually described them. In his monumental work *De Sedibus et Causis Morborum per Anatomen Indagatic* (The Seats and Causes of Diseases Investigated by Anatomy), which ran to three volumes and was published in 1681, Morgagni described these cysts in the pigeon and, incidentally, took some space to castigate the unfortunate servant who had accidentally got rid of the specimens before Morgagni had finished with them. Thus the little cysts became known as hydatids of Morgagni, another example of the paradox of a relatively unimportant observation attaining immortality while greater achievements gained no credit in history. But Morgagni's reputation needed no monument other than his published works. Even during his lifetime he was referred to as His Anatomic Majesty and the Prince of all European Anatomists.

The fimbriae of the Fallopian tubes caress the vital structures which, even more than the male contribution, lie at the heart of human reproduction, for it is the ovaries which carry the egg, the source of each new life. For centuries, however, there had been arguments and controversy about the origin of the human egg.

Aristotle had imposed his view on generations of doctor-philosophers that the human egg was produced in the womb by the interaction of menstrual blood with male semen. Galen, whose word was medical law in the Roman era, disputed this and claimed that the female testis produced a semen which mixed with the male semen in the uterus and that the resulting coagulum gave rise to the embryo. In many ways this was a paraphrase of the facts, but it was hundreds of years before the true mechanism was understood.

Vesalius and Fallopius saw that the ovary has small, fluid-filled cysts, but they had no idea that some of these vesicles carry the human egg. Fallopius's successor, Fabricius, gave the name ovary to the appropriate organ in the hen, and guessed its true function, but he did not transpose the concept to the human female. It is said that in the fifteenth century Ferrai d'Agrate, a Milanese, had indeed applied the name ovary to the human organ and assumed that it had the same purpose as the egg gland of the fowl. But no one can now dispute the role of the Dutch anatomist Regnier de Graaf in finally explaining and clearly describing the origin of the human female seed.

In 1672, Regnier de Graaf published *De Mulierum Organis Generatione Inservientibus*, a treatise on many aspects of female anatomy and disease. He devoted a whole chapter to the ovaries. In great detail he described the structure of the "female testis" of various animals and the human and described the vesicles which are now known as Graafian follicles, which he thought were whole eggs, rather than masses of cells with the actual ovum embedded within them. He also accurately explained the process of ovulation.

In 1666 and 1667, two physicians in Leyden, Jan Swammerdam and Johann van Horne, and a third in Denmark, Niels Stensen, had pursued the same theories about the ovary, but none of them had published his conclusions. A few weeks after the publication of *De Mulierum*, however, Swammerdam bitterly attacked de Graaf, accusing him of scientific plagiarism and claiming that it was he who had discovered the true function of the ovary. He insisted that he had been deprived of prior publication by delays and by the death of van Horne. It soon became known that Swammerdam had persuaded Stensen to withdraw from the contest.

Regnier de Graaf was deeply affected by Swammerdam's accusations and he went so far as to publish a rebuttal. It was said that his health suffered as a result and it may have been a factor in his untimely death at the age of thirty-two. A fitting epitaph for this unsavoury episode came from a professor of anatomy at another Dutch university who observed dryly that Swammerdam had "smeared the ovary not with honey, but with the most bitter gall."

It was more than one hundred and fifty years after de Graaf's death that the actual monocellular nature of the ovum within the depths of the Graafian follicle was understood. In 1827 Karl Ernst von Baer, an Estonian zoologist, dissected a follicle and finally took up the tiny cell on the point of his knife.

The stability of the female generative organs depends on a number of suspensory and bracing ligaments, and these in turn give rise to various pouches and culs-de-sac in the pelvis. One of the most distressing gynaecological conditions is prolapse of the uterus, due to a failure of this suspensory apparatus. The condition was mentioned in Egyptian papyri, but before there was adequate understanding of the supporting structures, the attempted cures for prolapse—inefficient, often dangerous, and sometimes horrific—included the application of leeches, burning, and the deliberate introduction of infection.

Alwin Mackendrodt, a nineteenth-century German gynaecologist, was responsible for the first real elucidation of the pelvic ligaments, and, therefore, an understanding of the nature of prolapse. His name became linked with the transverse cervical ligaments, also known as the cardinal ligaments, which are the major factor in uterine support. It was Mackendrodt's masterly account of pelvic mechanics that laid the foundation for the rational treatment of prolapse.

Also important in uterine support is the broad ligament which hangs from the uterine tubes and falls away from the T-shaped complex

Gabriele Fallopius was an Italian nobleman and the most illustrious anatomist of the sixteenth century. For eleven years he held the chairs of anatomy, surgery, and botany at Padua. In his short lifetime—he died of pleurisy in 1562 at the age of forty—he contradicted Galen fearlessly and made many important contributions to the study and understanding of human anatomy. In a scientific climate often notorious for competition and venomous recrimination, Fallopius's behaviour towards his colleagues was exemplary. He published only one work during his lifetime. This was *Observationes Anatomicae* which appeared in 1560.

Giovanni Battista Morgagni was for fifty-nine years professor of anatomy at Padua. Born in 1682, at the age of sixteen he began his medical studies with Valsalva at Bologna. While his work in anatomy was itself remarkable, he is best known for his monumental three-volume work on the causes of disease. Published in 1769, two years before his death, it is considered the foundation of the science of morbid anatomy. He made the study of disease a precise science, grasping the connection between disease and morbid pathology. Morgagni also wrote a book about Celsus, an archaeological study of Forli, and a biography of Valsalva.

Hieronymus Fabricius was born in Aquapendente in 1537. He began to study philosophy, but became a student of Fallopius in medicine at Padua and, in 1562, succeeded him as professor of anatomy. An outstanding lecturer, he taught and strongly influenced Casserio and Harvey. He also practised surgery and made contributions in the fields of embryology and comparative anatomy. Fabricius was not satisfied with the old Galenic systems and favoured the new stirrings in physiology. Unfortunately, he did not live to see the great scientific breakthrough. He died in 1619, nine years before Harvey's *De Motu Cordis* was published.

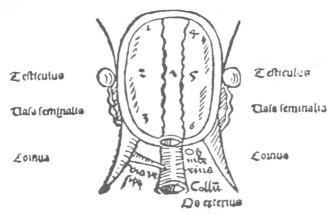

This hypothetical drawing of the uterus dates from the early sixteenth century, when most anatomists were still under the impression that the female had rudimentary testicles and a bicornate womb.

Regnier de Graaf made many notable contributions to the understanding of anatomy and physiology. Born in Schoonhaven in 1641, he went to France for his medical training and published his first monograph on digestive function while he was still a student. He qualified at Angers in 1667 and returned to the Netherlands two years later to practise medicine in Delft. He never held a university post, but he was a distinguished anatomist whose original research proved of great importance. He studied the pancreas and gall bladder using the experimental method of collecting the secretions of a dog. In 1668, in his book on the male generative organs, he described the structure of the prostate gland and seminal vesicles. In 1672, one year before his untimely death, he published his famous book on the female organs and discussed the changes in the body after the egg is fertilized. Ironically, the structures he

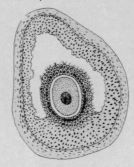

observed in the dissected ovary were identified incorrectly as ova. Much later, it was discovered that the Graafian follicle, in which the ovum matures, produces the hormone progesterone after ovulation to prepare the uterine lining for pregnancy.

formed by the uterus and the Fallopian tubes, rather as a cloak would fall from outstretched hands. In addition, two cord-like ligaments pass forwards and downwards from the points where the Fallopian tubes are attached to the womb. Known as the round ligaments, these are hidden beneath the fold of the broad ligament, emerging to end rather diffusely by passing through the inguinal canal in the groin.

In the developing foetus, the round ligaments carry with them a pouch of peritoneum, the lining membrane of the abdomen, rather like the finger of a glove. This pouch usually becomes obliterated before birth, but in some cases it survives into adult life. If it persists intact, it may allow access for a herniation of the intestine, but if it becomes sealed, then a closed cystic space may develop, and a swelling may appear in the groin or even in the labia of the girl or woman.

Physicians of antiquity were aware of this swelling and the first description is attributed to Aetius of Amida, who practised in Constantinople in about A.D. 550, and based his account upon the experience of a midwife. Two millennia later, Ambroise Paré, the great French pioneer of surgery in the sixteenth century, mentioned the surgical treatment of a young girl with the same lesion. In the following century, Jan Swammerdam drew a tongue of peritoneum around this ligament in an illustration which appeared in 1672 in his *Miraculum Naturae*. But it was Swammerdam's contemporary and compatriot Anton Nuck who, in 1691 in his book *Adenographia,* described the anatomy of the pouch in detail and was rewarded by having it named the canal of Nuck.

Nuck was investigating the causes of inguinal hernia in women, the condition in which a loop of small intestine becomes insinuated beneath the ligament demarcating the groin. Various fingers of peritoneum may creep beneath the inguinal ligament and Nuck made a meticulous study which included the canal that sometimes survives to ensheath the uterine round ligaments.

Below and to the rear of the uterine suspensory ligaments is a membranous cul-de-sac which sinks down between the vagina and the rectum and forms a sump for the lower end of the general peritoneal cavity containing the abdominal and pelvic organs. It is one of the best known pelvic structures, certainly in gynaecology and in lower abdominal surgery, and it is associated with James Douglas, a Scots obstetrician who was personal physician to Queen Caroline.

In 1730, Douglas published a book, *A Description of the Peritonaeum*. In this masterly account of the general cavity of the abdomen he devoted barely a paragraph to the pocket-shaped recess which now bears his name. Douglas did not use the word pouch and it therefore seems remarkable that so durable a term as the pouch of Douglas should have arisen from only a brief mention hidden among a great mass of description of other parts of the peritoneum.

Nothing is known of Douglas's origins and no portrait of him survives. But he was further commemorated in a poem *The Dunciad* by Alexander Pope, who wrote:

There all the Learn'd shall at the labour
stand
And Douglas lend his soft, obstetric
hand.

Just as the anatomy of the suspensory apparatus of the female organs has direct clinical applications, so does the nerve supply, especially to the uterus. Because the reception of pain and the transmission of stimuli to muscle are so relevant to childbirth, an understanding of uterine innervation was essential to advances in obstetrics. Yet this knowledge came relatively late and was accompanied by professional antagonisms at least equal to the trouble between de Graaf and Swammerdam.

De Graaf had made tentative forays into this field as did such other well-known anatomists as John Hunter and Eustachius, but it was a Scotsman, Robert Lee, who first published an accurate and detailed account of the pelvic nerves. In 1841, Lee, an obstetrician, delivered a superb paper to the Royal Society. Later published and beautifully illustrated, this paper gave a definitive description of the way the sympathetic and sacral nerves supply the uterus. They pass from the meshed plexus of nerve fibres, over the division of the aorta, down to form the two hypogastric nerves. From these are formed the nerve centres of the pelvic organs, the two great utero-cervical ganglia, which are condensations of nerve fibres and cells, each situated at the side of the womb at the level of its neck. Also called the hypogastric ganglia, these are the nerves which supply the uterus, vagina, bladder, and rectum and are, therefore, all-important in the sensory and motor functions of the pelvic organs.

Lee described the anatomy in minute detail and showed how the nervous tissues are able to accommodate the great changes that take place during pregnancy—and return to normal afterwards. The vital utero-cervical ganglion later became known as Lee's ganglion, but the establishment of Lee's claim to the eponym was clouded by an almost diabolical conspiracy to defraud him of the honour.

Until 1841, when he delivered his paper, Robert Lee led an unremarkable professional life. Born in Melrose in 1793, he graduated in medicine from the University of Edinburgh

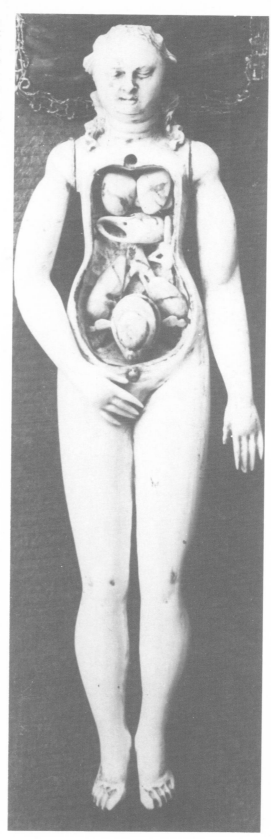

This little ivory figure, with removable parts, representing a pregnant woman, was probably used in the sixteenth century for the instruction of midwives.

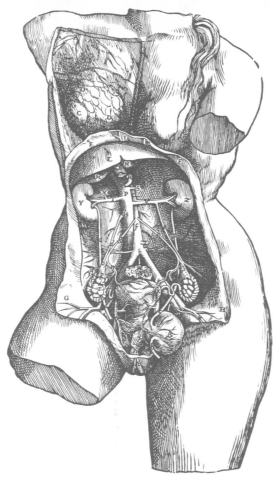

One dissection, of a woman who hanged herself, provided the basis for this elegant, but not entirely correct drawing in Vesalius's *De Fabrica*. There are a number of anatomical misconceptions based mostly on persistent though mythical analogies with male anatomy. On the other hand, the brilliant draughtsmanship is a great advance over the charming, but crude, woodcuts of the Middle Ages.

Astley Paston Cooper was one of the most skilful and popular surgeons of his time. Born in Norfolk in 1768, the son of a clergyman, he went to London at the age of sixteen and studied under John Hunter. In 1800, Cooper succeeded his uncle as surgeon at Guy's Hospital and, from 1813, was professor of comparative anatomy at the Royal College of Surgeons. He was also a prolific writer and published a number of books in the course of his long and productive life. Cooper, who received a baronetcy in 1820, died in 1841 and was buried in the chapel of Guy's Hospital. A memorial to him was placed in St. Paul's Cathedral.

at the age of twenty. He soon went to London, as did so many aspiring Scots obstetricians. Lee studied anatomy first in London and then in Paris before returning to London to practise and to specialize in obstetrics. He took the opportunity to return to Scotland in 1834, when he was offered the chair of midwifery at Glasgow. But London again attracted him and he returned to become professor of obstetrics at St. George's Hospital, where he remained until the end of his professional career when he was seventy-three.

It was during his many years at St. George's that Robert Lee carried out his anatomical dissections of the pelvic nerves, but his work was dogged by jealousy and hostility. Dr. Snow Beck, one of his former students, became locked in competition with his old teacher for the Royal Medal in Physiology, awarded by the Royal Society. To further his own chances, Beck clandestinely damaged Lee's dissected specimens that were kept in the museum of the Royal College of Surgeons. Beck had friends on the committee of the Royal Society that was responsible for awarding the medal. They helped him by holding unconstitutional meetings, of which they suppressed records, even erasing a minute from the committee's journal.

The sordid affair led to angry and sustained correspondence in the *Lancet*. The editor of that illustrious medical journal came out on Lee's side, however, and eventually helped to vindicate him and to confirm his claim as the true explorer of the pelvic nerves and Lee's ganglion.

Although Lee had produced a masterly account of uterine innervation, even greater detail was achieved twenty-six years later by Ferdinand Frankenhauser, an equally meticulous German professor of obstetrics. He published a treatise in 1867, which not only expounded the minute anatomy of the pelvic plexuses and ganglia, but related function to structure in a way which was directly applicable to the practice of obstetrics. The great uterocervical ganglion, upon which Robert Lee's name secured a rather tenuous hold, was then known as Frankenhauser's ganglion, leading to the not uncommon situation of two eponyms being attached to the same structure.

Frankenhauser was born in 1832, but nothing more is known of his origins or education. He first appears in Jena and began making a name for himself in the middle of the nineteenth century with a paper on the relationship of infant size to the bone structure and fecundity of the mother. He became embroiled in a controversy concerning the theory that the sex of an unborn infant could be predicted from its heart rate while still in the womb, but then moved on to firmer ground with a lasting interest in the innervation of the uterus.

Frankenhauser's work on this subject was done at Jena, but he was invited to Switzerland to take the chair of obstetrics at Zürich where he spent the rest of his professional life. Although his name is associated with the great ganglion, he made an impressive reputation in his later years as an excellent practical obstetrician and gynaecologist. He reduced maternal mortality in Zürich to the lowest level yet attained and also developed a surgical technique for the removal of large uterine fibroid tumours through the vagina, to avoid the high rate of abdominal infection common in those days of inadequate antisepsis.

Even in the ancient, almost diagrammatic representations of the pregnant female body the external organs were rarely depicted. Although it was uncommon, occasionally a body was drawn with breasts. Possibly because there were fewer female than male cadavers available for dissection, it was not until the nineteenth century that the muscles and ligaments which give the breast its fullness and form were fully investigated.

Around the nipple the circle of darker skin is called the areola, a Latin diminutive of the word for area. It is under this and the nipple itself that the various layers of muscle lie. Some circular, some radiating outwards, all these muscle layers are involved in the firming and erection of the nipple in response to a baby's sucking or to erotic stimulation.

The radial muscle of the nipple was named Meyerholtz's muscle by the famous nineteenth-century anatomist Freidrich Henle, who was the director of the Anatomy Institute in Göttingen, Germany, where it seems that Meyerholtz worked at the time he described the muscle. Mysteriously, no publication can be traced under his name, although in books published much later two well-known French anatomists also ascribed the muscle to Meyerholtz.

The circular muscle of the nipple is known as Sappey's plexus. It was described by Marie Phillipe Constant Sappey, a French anatomist who worked in Paris in the nineteenth century and was well known for the meticulous accuracy and detail of his work. The thin sheet of fibres which runs under the nipple and its surrounding areola is known as Marcacci's muscle after Arturo Marcacci, a nineteenth-century Italian anatomist.

A network of fine fibrous strands intersects the soft glandular structure of the breast, to keep its shape and to form a framework within which the pliable gland tissue can be supported. These ligaments, known as Cooper's suspensory ligaments, commemorate Sir Astley Paston Cooper, the outstanding English surgeon who described them in 1845.

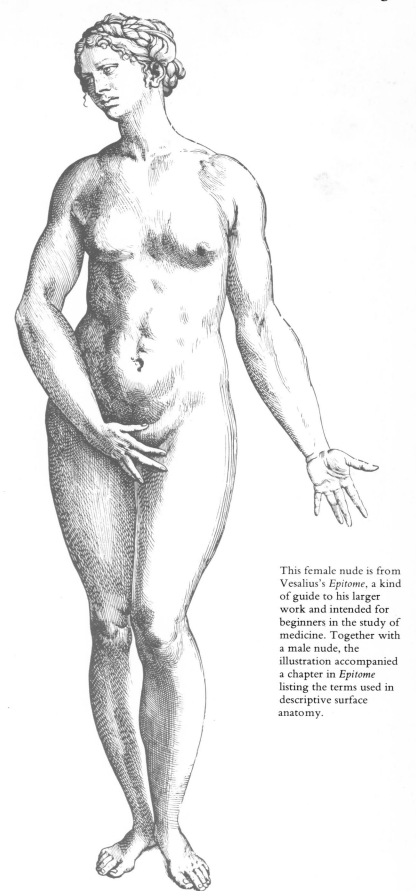

This female nude is from Vesalius's *Epitome*, a kind of guide to his larger work and intended for beginners in the study of medicine. Together with a male nude, the illustration accompanied a chapter in *Epitome* listing the terms used in descriptive surface anatomy.

Leonardo's studies of the womb and the foetus were the first to accurately depict pregnancy and the unborn child.

Pregnancy and the Unborn Child

The process of the generation of new human beings has intrigued men and women for millennia, yet until the last few centuries there was little real knowledge of the way the child was conceived or how it developed. Childbirth was a more practical matter and midwifery had been practised since civilization began. Indeed, midwives and the long-established birthstool are mentioned in the Old Testament.

In the fourth century B.C., Diocles, who came from Carystus in Euboea and practised medicine in Athens, developed a theory of embryology that was nearer the truth than the later Aristotelian ideas. Diocles believed that the new child came from a fusion of the seed of both sexes. He said that he had examined a foetus of twenty-seven days since conception and found in it traces of the head and spinal column. He claimed he was able to identify an embryo at forty days as human. Most of his dissections were of animals, but he must have seen some human material and although his book *On*

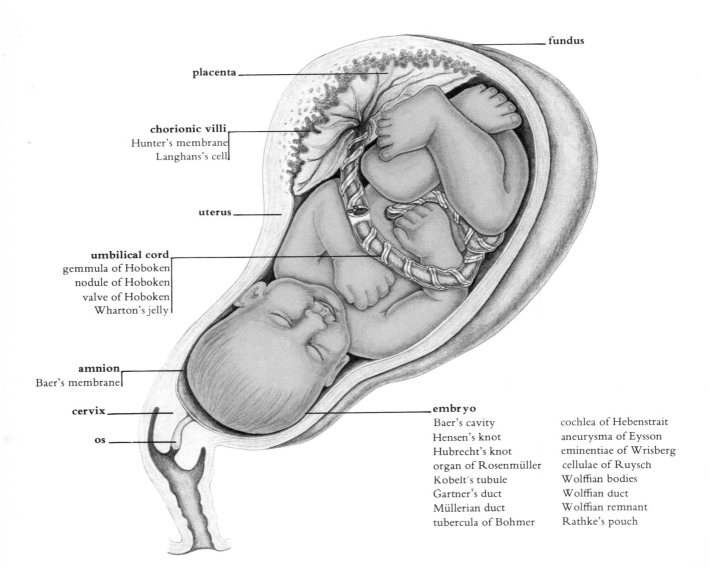

fundus

placenta

chorionic villi
Hunter's membrane
Langhans's cell

uterus

umbilical cord
gemmula of Hoboken
nodule of Hoboken
valve of Hoboken
Wharton's jelly

amnion
Baer's membrane

cervix

os

embryo
Baer's cavity
Hensen's knot
Hubrecht's knot
organ of Rosenmüller
Kobelt's tubule
Gartner's duct
Müllerian duct
tubercula of Bohmer

cochlea of Hebenstrait
aneurysma of Eysson
eminentiae of Wrisberg
cellulae of Ruysch
Wolffian bodies
Wolffian duct
Wolffian remnant
Rathke's pouch

Anatomy has not survived, he may well be one of the first recorded embryologists.

Polybus, the son-in-law of Hippocrates, was slightly earlier than Diocles and may have influenced him, for Polybus also wrote about generation. He stated that acquired characteristics are inherited and that the embryo receives nourishment through the umbilical cord.

Aristotle was an indefatigable researcher into all things biological. Hen's eggs provided most of the basis for his embryological studies. He wrote the first major book on embryology, which covered many types of creature, including fish and crustacea. As far as the human was concerned, although Aristotle did no practical dissections, he developed the theory that the embryo was created from female menstrual blood. He denied that there was any female "seed" and that the father contributed anything physical to the new child. Aristotle knew of the male semen and the generative apparatus, nevertheless he believed that the life force, or soul, of the foetus came entirely from the father and that the mother merely provided the physical material for the growth of the new body.

In Aeschylus's play *Eumenides,* this concept, which may be of Egyptian origin, is evident when Apollo says, "The mother of what is called her child is no parent of it, but nurse only of the young life that is sown in her. The parent is the male and she but a stranger, a friend, who if fate spares his plant, preserves it till it puts forth."

This concept of Aristotle's that the physical seed of the father was not necessary for generation permitted the way to remain open for the theory of parthenogenesis, or virgin birth. This in turn fitted in with the dogmatic needs of the early Christian Church in the matter of the immaculate conception, hence their ready acceptance of the views of Aristotle, who was otherwise pagan to them.

In the third century B.C., Herophilus wrote a book for midwives. He must have dissected human embryos and gave a fair account of the internal structure of the umbilical cord.

Much later, Soranus of Ephesus, another doctor with an interest in pregnancy who was educated at Alexandria, went to Rome to practise in A.D. 100. He wrote a book on obstetrics and gynaecology which was rediscovered only in 1838, although parts of it were used by Oribasius, the fourth-century Greek writer, and also in a sixth-century Latin tract entitled *Muscio.*

This work of Soranus's is interesting since it contained the first known, fairly accurate medical illustration of the uterus. It names the fundus as well as the cervix and the os at the lower end. It was a better representation than later drawings right through the Renaissance.

Casserius prepared this romanticized depiction of the foetus in the womb which was first published in 1626 in Adriaan van der Spieghel's book *De Formato Foetu.*

Soranus also made a diagram showing the complete anatomy of the pregnant woman, on which some medieval drawings may well have been based. Although his full text was lost until the nineteenth century, the Latin *Muscio* was widely used in the Middle Ages.

Soon after Soranus, the great figure of Galen emerged. Embryology was well represented among his legion of medical interests, although he did not make any great contributions to the subject. He must be credited, however, for diverging from Aristotle's view that the male was all-important, but that neither mother nor father need contribute physical seed to the process of generation of a new body.

Galen claimed, correctly, that the testicles correspond to the ovaries and that both the testicles and the ovaries secrete a substance that forms the embryo. He perpetuated the long-standing myth, however, of the bicornate human uterus. Galen believed that in the embryo the liver was congealed from the blood and was the first organ to be formed. The brain was then made from the seed, followed by the heart which was congealed from the blood. These theories, too, persisted for a millennium or more.

During the long silence between Galen and the Renaissance, there were a few murmurs in the field of generation. One interesting voice was Albertus Magnus, a Dominican friar from Cologne. Born in 1193, Albertus was a teacher of St. Thomas Aquinas. At one time Bishop of Regensburg, he devoted most of his attention to biology and was a devout, although not

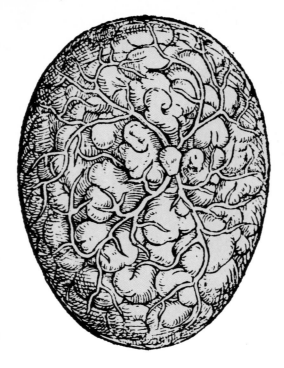 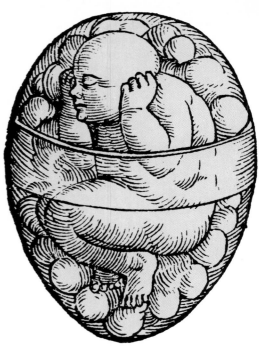

Aristotle believed that sperm mixed together with menstrual blood to make the human embryo. A handbook for midwives, published in 1554, illustrated this belief. The amalgam was said to produce first a blood system and then the bones, muscles, and skin.

uncritical, follower of Aristotle. Albertus wrote twenty-six books about animals, seven of which contained writings about embryology. Many of his arguments were vague and complex, but there are flashes of illumination. Albertus believed, for example, that there was definitely a female seed, which combined with the male seed to form a coagulum, added to which was menstrual blood that made the new human.

Like Aristotle and other earlier embryologists, Albertus found the chicken's egg the most convenient for studying the progress of foetal development and much of his writing concerns this. He contradicted the well-established theory held since Aristotle's time that in fowls, males come from sharp-ended eggs and females from the rounder, blunter eggs. His writings, including *De Secretis Mulierum*, were much copied until the beginning of the seventeenth century and were also the basis of *Aristotle's Masterpieces,* a small book which is still in print today.

The scientific study which began during the Renaissance was given impetus by a number of famous anatomists, including Fabricius and Leonardo. Unfortunately, the work of Leonardo da Vinci was lost at that time and appeared again so recently that it never influenced the mainstream of anatomical evolution. Leonardo's work on embryology, which was in the third volume of his notebooks, was only published in 1911, when Norwegian scholars produced a facsimile edition.

Leonardo dissected the human pregnant uterus and made suberb drawings of its contents. He had a good working knowledge of the process of childbirth and recognized the various parts and membranes of the pregnant womb. The amnion and chorion of the membranous bag were described by him, although he drew the placenta as a collection of masses, the cotyledons characteristic of many animals, rather than the single large afterbirth of the human. Although he reverted to the Aristotelian theory that long eggs gave rise to male chicks, he discerned other matters with a remarkable degree of accuracy.

The discovery of the circulation of the blood was still almost one hundred and fifty years in the future, yet Leonardo asserted that the foetal blood system was not continuous with that of the mother. His acute appreciation of physics also led him to say that the child lay in a bag of water in the womb, because heavy things weigh less in water, and that the water distributes the whole weight of the baby evenly over the body and the sides of the womb.

Leonardo recognized, too, that the baby does not breathe in the womb, "otherwise it would drown." It did not need to, he said, because it gets its nourishment from the mother through the umbilical cord. He was convinced that the foetus urinates into the womb fluid and denied that it can cry, saying that the noise that some women allege is due to the foetus passing flatus. He also made some quantitative calculations on the rate of growth of the child in utero. In many ways Leonardo da Vinci was the father of

This illustration of the foetus in the womb, from one of Leonardo's anatomical notebooks, has the commentary written in the mirror script that he invariably used.

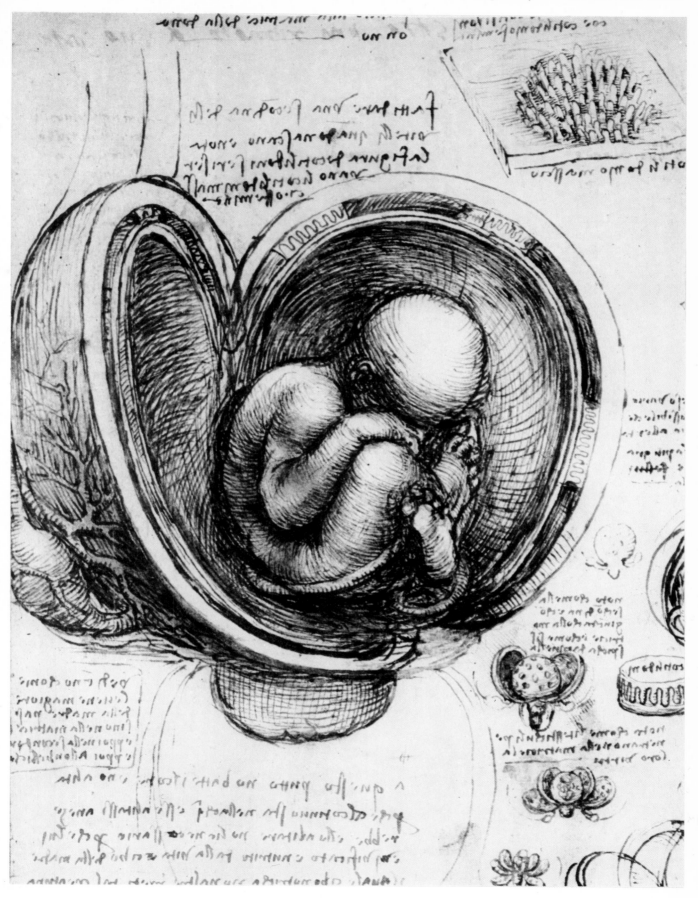

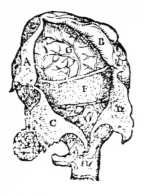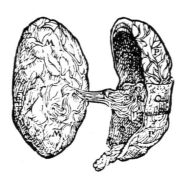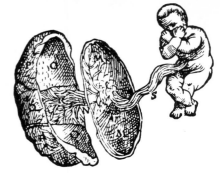

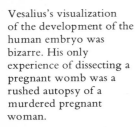

Vesalius's visualization of the development of the human embryo was bizarre. His only experience of dissecting a pregnant womb was a rushed autopsy of a murdered pregnant woman.

scientific embryology, although no one knew it until centuries later.

Many anatomists, including Volcher Coiter, a Dutchman who, in the sixteenth century, studied embryology in the chick, also took a marked interest in generation, which was largely ignored by Vesalius and some of the other major anatomy teachers and writers. In 1571, when he was a professor at Bologna, Julius Caesar Arantius wrote a book *The Human Foetus*. This was the first useful account of pregnancy to appear in print and the first to show definitely that the human placenta is a single mass and is not subdivided into cotyledons. Arantius gave an excellent description of the anatomy of the human foetus, especially of the heart and the foetal structures of ductus arteriosus and foramen ovale, which are part of the bypass system which ceases to function when the baby begins to breathe air.

Charm is substituted for anatomical accuracy in this sixteenth-century woodcut showing twins in the womb. There is no apparent impediment to the twins' lounging at full height in the womb in a fraternal embrace.

Fabricius of Aquapendente was particularly interested in the formation of the foetal heart and of the changes that take place at birth. He wrote a well-regarded book on embryology, *On the Development of the Eggs of Birds*. It was a landmark in the history of embryology and was later used by Harvey in his own research into the subject of generation. Charles Singer, the foremost historian of anatomy, wrote of Fabricius, "He carried the subject far beyond where Coiter left it and elevated Embryology at one bound into an independent science." Fabricius had to rely entirely on his unaided eyesight for his work, but nevertheless the book contains excellent illustrations, although the early stages, before the sixth day of development of the egg, had to await the microscope in the hands of Malpighi.

Fabricius wrote a second book, *On the Formed Foetus*, which concerned the more mature embryo of many species including humans. This was published in 1604 along with his other book. There is no doubt that the writings of Fabricius were by far the most advanced works on embryology of his time.

It was the work of William Harvey that marked the modern period of embryology. Harvey spent the latter part of his professional life studying generation and the results are almost as outstanding as his cardio-vascular achievement. Just as Fabricius, his teacher at Padua, stimulated Harvey's interest in veins and blood vessels, so the Italian master endowed him with a lasting curiosity about the development of the embryo.

After the defeat of the Royalists in the Civil War, Harvey retired to Oxford and there did much of the work on chickens' eggs, which led to the publication, in 1651, of his second great book *De Generatione Animalium*. In addition to its embryological content, it was the first book in English to be concerned with practical midwifery; one chapter, *De Partu*, is devoted to obstetrics.

Harvey's statements about embryology were typically firm and careful, as well as novel. He gave a new direction to thought on the development of the foetus, both in the chick and in some mammals. He established two important facts. The first is that all animals come from eggs, or ova, and from nowhere else. The second is that all the parts of an embryo arise by new formation from the original ovum and not by mere enlargement of a tiny preformed animal, as was formerly believed. This vital principle, known as epigenesis, was hotly debated for the next century and longer.

The only instrument that Harvey had with which to examine the embryo was a magnifying glass and so he was unable to study the earliest stages of development. This makes his achievement all the more remarkable, for he made faithful descriptions of the stages of development after the first thirty-six hours.

Marcello Malpighi, with access to a microscope, continued Harvey's work on the embryo, just as he had on the circulation. In 1672, Malpighi was able to communicate his findings to the Royal Society in London. In two tracts on the developing chicken embryo, he traced the formation of the chick back to the earliest stages. His work was not bettered until the nineteenth century.

In the seventeenth century, Walter Needham, Regnier de Graaf, and Jan Swammerdam, were among the anatomists who helped to show the similarities in the development of different animals. De Graaf, Nicholas Stensen, and Johann van Horne were particularly important in the discovery that the ovarian follicle containing the ovum in humans and other mammals is the prime stage that bursts from the ovary. This work was attended by a notorious con-

troversy between Swammerdam and de Graaf over priority of discovery.

Walter Needham, a Cambridge physician, went to Oxford in 1660 to study with the remarkable group of men that included Christopher Wren, Richard Lower, and Thomas Willis. Seven years later, Needham published his book *The Formation of the Foetus* in which he recorded the first chemical experiments on developing embryos, as well as the first directions for dissecting the foetus. His chemistry was crude because of the poor facilities and limited knowledge of the time, but the book was a step in the right direction and Needham made genuine advances in the knowledge of the foetal circulation.

Two of the Oxford group, John Mayow and Robert Boyle, even more famous for their work on adult respiration and circulation, studied the respiration of the foetus and published in the late seventeenth century. In 1674, John Mayow wrote his book, which was almost forgotten until 1790, when his work on respiration was praised as being equal to that of William Harvey. Boyle's well-known book on chemistry, the *Sceptical Chymist*, which appeared in 1661, also contained important references to the respiration of the foetus.

In the eighteenth century, research continued to accelerate. The work of Hermann Boerhaave and Albrecht von Haller dominated the first half of the century. Boerhaave, an outstanding Dutch physician and famous teacher, opposed Harvey's views about epigenesis, the power of cells to differentiate into all types of tissue purely by variable development. He believed

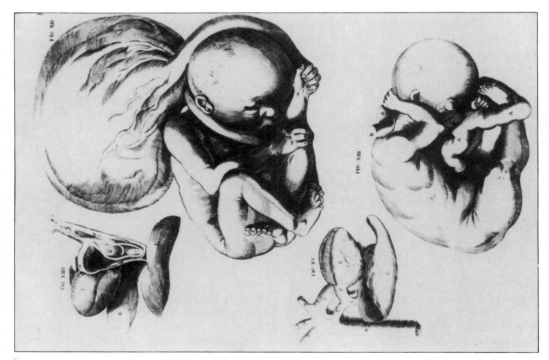

These fine drawings of the human foetus were done by Fabricius and appeared in his treatise *On the Formed Foetus*. The work contained the best illustrations up to its time of the human foetus and membranes and of the human placenta.

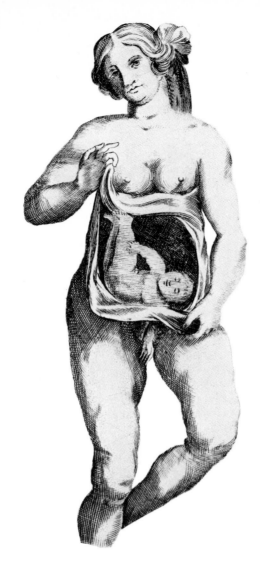

This drawing, captioned "The Successful Extraction of a Dead Foetus in an Awkward Position," appeared in a German book published in 1680. The only realistic element is that during birth the baby's hand could, just possibly, emerge first. The instrument, left, looks more suitable for crocheting than for performing a Caesarian section.

that all the organs were present in the germ of the embryo—a difficult theory to sustain once microscopic evidence was available to the contrary.

His pupil Albrecht von Haller rejected that theory and correctly supported Harvey's theory of epigenesis. He wrote voluminously on embryology in the middle and second half of the century. One of his major contributions was a calculation of the growth rate of the embryo, which involved complex mathematics, but was basically correct.

In 1755, William Hunter, the Scots anatomist and surgeon, published a book entitled *The Anatomy of the Human Gravid Uterus,* in which he called the lining of the maternal womb the decidua. It was later named Hunter's membrane. Although not unlike the chorionic membranes or placenta, the decidua is not part of the foetus or its adnexae. It is the lush vascular lining which develops only in response to the hormone changes caused by pregnancy. This pro-

vides the soil in which the placenta can take root to derive the blood constituents which feed the foetus.

Another man remembered for the membranes in the pregnant womb was Theodor Langhans, a pathologist in Berne, Switzerland. Langhans, who lived from 1839 until 1915, is best known eponymously in pathology for his Langhans' cells, which are giant cells typical of tuberculosis, but in embryology, he described other cells, those which line the finger-like processes of the placental surface and were also called Langhans' cells. The very large surface area of the chorionic villi, the fingers that mesh with the decidua of the uterine lining, have two layers of cells and Langhans described the inner layer.

In the nineteenth century, one of the outstanding embryologists was the Estonian Karl Ernst von Baer, who discovered the ovum cell itself, as distinct from the ball of surrounding cells called the follicle, which was previously thought to be the fundamental structure. Von Baer's work began about 1819 at Königsberg. In St. Petersburg in 1827, working with the Latvian Heinrich Pandar, he made his discovery of the ovum and in the same year began his series of publications which were classics in the world of embryology. His name was eponymously attached to the chorionic membrane, the lining of the pregnant womb that covers the foetus. A zoologist, von Baer examined many species and confirmed that the ovum was broadly similar in all of them since it was a totally potent cell which is able to give rise to all the body tissues.

Inside the actual ovum, which is one of the largest single cells in the mammalian body, is a zone of radiating fibrils visible only under a microscope. This was seen in 1842 by a German anatomist, Theodor Bischoff, and came to be called Bischoff's corona. The yolk nucleus of the ovum was described by Eduard Balbiani in 1893 and was called Balbiani's body. Balbiani, although born in Haiti, studied medicine in France and became professor of comparative embryology at the College de France in Paris.

Edward Pfluger, the leading figure among German physiologists in the second half of the nineteenth century, described cords or tubules of cells which eventually give rise to the ova which lie in de Graaf's follicles and so represent part of the chain of immortality; even the young female foetus is preparing to potentiate the human species. When a baby girl is born, she has within her ovaries between forty thousand and three hundred thousand ova, of which she will shed a maximum of about five hundred. The formation of ova is, therefore, a ripening process rather than one of manufacture.

Pfluger had a long professional life which spanned more than sixty years. He wrote prolifically, publishing hundreds of papers, including the important monograph dealing with the forerunners of the ovarian germinal cells, forever known as Pfluger's tubules.

Until there were chemical, biological, or radiological tests for pregnancy, physicians had their own methods of detecting the first signs of pregnancy, most of which involved the vagina or uterus. William Featherstone Montgomery, one of Ireland's most famous obstetricians in the early part of the nineteenth century, was an ardent champion of the breast as the harbinger of pregnancy. In his masterly book *An Exposition of the Signs and Symptoms of Pregnancy,* published in London in 1837, he described the mammary changes in great detail, commenting, "I wish to repeat that my confidence in the condition of the areola as a diagnostic mark of pregnancy is not only unabated, but very much increased by further observation." Montgomery was not the first to notice the small protruberances on the areola, the ring around the nipple which darkens during pregnancy. They had been noticed by Morgagni, whose name was for a time linked with them, but Montgomery completely upstaged him a century later and they became known as Montgomery's tubercles.

Yet another anatomist had described the tubercles on the nipple and Montgomery acknowledged this in his book. This was Rosderer of Göttingen, who had briefly described them in 1753: "The areola enlarges to a greater diameter and contains small protruberances, as if covered all over with tiny nipples."

Montgomery earned his claim to eponymous fame with a much more detailed description and linked the nodules firmly with the onset of pregnancy, claiming that they were a much more reliable sign than mere darkening of the areola, which had been relied upon by many other writers on the subject.

The length of the gestation period—how long it is from conception to delivery—is a calculation which exercised minds great and small for centuries. William Harvey had a firm religious approach to the subject. He flatly stated that the period of pregnancy was that of Jesus Christ and calculated it from the festival of the Annunciation in March to Christmas Day, which works out at two hundred and seventy-five days.

Franz Carl Nägele, one of the greatest obstetricians of the early nineteenth century, consolidated a simple rule, which was to add seven days to the date of the first day of the last menstrual period and then count back three months. This calculation had been used long before Nägele was even born, but for some odd reason all obstetric textbooks from his time attributed it to him and called it Nä-

William Hunter, who was born in Scotland in 1718, devoted himself to obstetrics and established it as a branch of medicine. He began to study at Glasgow, but went to London as a pupil in surgery. In 1747, Hunter travelled in France and observed medical students dissecting corpses. When he returned to London he opened his own school of anatomy where he allowed his students to perform autopsies. After the Royal Academy was founded, he was made professor of anatomy and was later elected president. Hunter, who served as physician and obstetrician to Queen Charlotte, died—probably from overwork—in 1783.

gele's rule. Actually he never claimed any originality for it, but merely quoted it in Latin from a book written by Hermann Boerhaave and published in 1744. Boerhaave himself referred to the calculation as being derived "from the numerous observations in France."

Today it is generally accepted that labour begins, on average, about two hundred and eighty days after the first day of the mother's last menstrual period.

The tiny cavity that develops in the first stages of division of the ovum, when it is a mere ball of cells, is called the blastocoele segmentation cavity, or Baer's cavity, commemorating Karl Ernst von Baer who described it in 1827, the same year as his discovery of the ovum.

When the early embryo develops further, it begins to form a central thickening, the primitive streak, which marks the site of the eventual spinal column. The first signs of this are known either as Hensen's knot or Hubrecht's knot. Both men were associated with this important stage of development, but Viktor Hensen, a professor of physiology at Kiel University, was some twenty-three years ahead of his competitor. Hensen did a great deal of work both on embryology and the fine structure of the sense organs, describing his knot in 1882. Ambrosius Hubrecht, a Dutch zoologist, became keeper of the zoological museum in Leyden and later was professor of anatomy at Utrecht; it was 1905 before he described the primitive streak.

In the embryo the primitive masses of cells in the abdomen which will give rise to much of the renal and genital organs are named the Wolffian bodies and the duct that becomes part of the excretory system is called the Wolffian duct.

The illustration above is part of a bas-relief in the Temple of Esneh and depicts the birth of Cleopatra's child. The size of the child is an indication of its royal parentage. Cleopatra is shown in the squatting position assumed by women in childbirth for centuries, since it permits gravity to assist the birth.

In ancient times there were many, rather barbaric, methods of hastening labour. Those depicted to the right and below were used by the Greeks even in the time of Hippocrates. On the right, the woman is repeatedly lifted and dropped on a couch. With the other even rougher method, the woman was tied to a couch, the couch was turned upright, and continually pounded on a bundle of faggots.

The curiously contorted woman below has taken a position presumed to aid labour, as recommended by Scipione Mercurio, a famous sixteenth-century Italian physician and obstetrician, who wrote a book called *The Godmother or the Midwife*.

The illustration below, from the book by the Italian obstetrician Mercurio, shows a birth by Caesarian section. Although the book was called *The Godmother or the Midwife*, the woman has only male attendants, since women were not even permitted to assist at surgery.

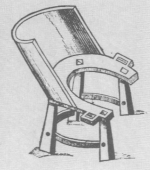

Caspar Wolff was German and spent much time in Berlin, where he was born and studied before gaining a post as a professor of anatomy at Breslau. He later returned to Berlin, but encountered so much opposition and professional jealousy that he went to St. Petersburg where he lived and taught until his death in 1794.

Wolff reported the bodies and the highly complex three-dimensional process by which they develop in 1759, when he was only twenty-six, in his doctoral thesis *Theoria Generationis,* which is recognized as the initiation of modern embryology. The basis of Wolff's work was the demonstration of granules, actually cells in the chicken egg, which grow into the organs of the body of the chick, even though the original microscopic structure bears no resemblance to the final organs and tissues. His work showed beyond all doubt that no preformed mini body is present in the egg and that cells have the power to differentiate into all types of tissues and organs purely by variable development. Although this seems so obvious today, it was an important discovery in the eighteenth century when the possibility of epigenesis was by no means universally accepted.

Wolff's *Theory of Generation* is one of the most important works in the history of the subject and established embryology on a secure scientific footing. Whenever the development of the genital system in either sex or the genesis of the excretory system is discussed, the name Wolff is an inevitable part of the dialogue.

The Wolffian duct and bodies were named by Martin Heinrich Rathke, a German anatomist who is himself well remembered by an eponym—Rathke's pouch, a diverticulum which gives rise to the anterior part of the pituitary gland. Rathke, who was an anatomist, physiologist, pathologist, and zoologist, made important contributions to the study of embryology. He described his pouch in 1861 in his textbook of embryology.

Although the Wolffian body is essentially an embryological structure in the foetal female, a remnant of it persists in the ovary until after birth. This was discovered by Johann Christian Rosenmüller, another patient and meticulous German investigator who was professor of anatomy and surgery in Leipzig. In 1802, he described the organ of Rosenmüller in a pamphlet and although more modern terminology calls the Wolffian remnant the paraovium, the eponymous title is still well known. Within the organ of Rosenmüller, the little tubules which make up the body are named after George Kobelt who described them in 1847.

Yet another name is applied to the remnant of the Wolffian duct which leads out of the body. Anatomically known as the mesonephric duct, it breaks up and partially vanishes as the

This woodcut below, from the works of a Dutch physician in 1681, shows the ludicrous lengths to which female modesty and medical prudery could go. The physician is attending a birth, but the patient is shrouded in a sheet and the doctor must carry out his manipulations blindly.

The obstetrical chair upon which women sat during childbirth is mentioned in the Old Testament and was sensible, since it enabled gravity to assist birth. In the illustration left, the pregnant woman is being modestly assisted by a midwife. The woodcut appeared in a sixteenth-century book *The Garden of Roses for Pregnant Women.*

foetus grows, but remnants persist in up to 40 per cent of newborn infants, and in 1 per cent of adult females cells belonging to the Wolffian duct can be identified microscopically.

Although the astute Malpighi noted the mesonephric duct in the cow in the seventeenth century, it was Herman Treschow Gartner, a Danish anatomist, who described it in detail. Gartner, who was born in the West Indies but moved to Copenhagen in childhood, entered the university to study medicine in 1803. He had a varied professional life which included military service in the Norwegian Army and post-graduate study in London and Edinburgh. He studied the surgical anatomy of hernias, but his anatomical research resulted in a description of the mesonephric duct in 1822, which immediately caused his name to be attached to the structure. It has been known as Gartner's duct ever since, although sometimes Malpighi's name is added.

Yet another duct is celebrated with a name. Johannes Müller, son of a cobbler from Coblenz on the Rhine, was a genius and polymath who found time for poetry and literature as well as original medical research. His name has gained medical immortality through being linked with the developing female generative organs. The origins from the paired Wolffian bodies were also paired. Although the ovaries and tubes remain double, partial fusion occurs to produce the uterus and vagina. In 1830, Müller was able to correlate much of the recent research of the early nineteenth century and create a coherent picture of the development of the mammalian womb. The precursors of the tubes and uterus have consequently become known as the Müllerian ducts.

It was inevitable that the visible link between the pregnant mother and the unborn foetus should receive considerable attention from embryologists. The umbilical cord is a pale, slippery rope, consisting almost entirely of the blood vessels that carry blood to and from the placenta. It is derived from the products of fusion of the egg and the sperm, and therefore belongs, as does the placenta, to the foetus and not to the mother.

Within the umbilical cord the umbilical veins and the single artery have an anatomical configuration unlike any other blood vessels. They have internal folds, ridges, and spirals long thought to have a valvular action, although this has not been confirmed by modern research. It may be that they are related to the sudden need for contraction of the vessels at the time of birth or they may even be artefacts seen only on examination of the defunct cord.

They were observed by a number of alert anatomists in the seventeenth century, and many eponyms have been applied to the bulges between each fold. These include the tubercula of Bohmer, the cochleae of Hebenstrait, the aneurysma of Eysson, the eminentiae of Wrisberg, and the cellulae of Ruysch. In addition there are the gemmulae of Hoboken, the nodules of Hoboken, and the folds themselves are best known as the valves of Hoboken. The name of Nicholas Hoboken is associated no less than three times with structures that form the inner lining of the umbilical vessels. The anatomist Hoboken was born in Holland in 1632. He lived and studied in Utrecht and acquired a doctorate in philosophy there before he graduated in medicine. Hoboken is best known for his work on the placenta and the umbilical cord. He published a book *Anatomis Secundinae Humanae* in 1669, which he beautifully illustrated himself, including drawings of Hoboken's valves, folds, and nodes.

Between the corded veins and artery of the umbilical cord, the ground substance that fills the space between the overlying covering membrane is called Wharton's jelly. This rather glutinous, colourless substance acts as a buffer against impact and, since it is semifluid and mobile, it can absorb the pulsing of the rapid foetal heartbeat and accommodate the sudden shrinking of the umbilical vessels at birth.

This remarkable substance was first described in detail by Thomas Wharton, the seventeenth-century London physician who was one of the few doctors to stay behind to treat the sick during the great outbreak of plague that ravished London in 1665. In his scientific work, Wharton has been remembered for his anatomical writings, rather than for any clinical research. His main work was his book *Adenographia*, published in 1656, which was an account of all the known glands of the body.

An interesting sidelight on Thomas Wharton can be found in the famous book *The Compleat*

Johannes Müller, the shoemaker's son, was one of the outstanding teachers of medicine of his time. Born in 1801, he studied at the University of Bonn. As professor of anatomy and physiology in Berlin, he taught his students—who included Henle, Virchow, von Kölliker, von Helmholtz, and Schwann—to apply the scientific method of precise experimentation and observation to problems in medicine. Müller greatly advanced the study of anatomy, embryology, and physiology, by himself and through the pupils he inspired. A prolific worker, he was subject to bouts of depression, and his death in 1858 may have been a suicide.

Angler by Isaac Walton. In this volume of 1652, subtitled "The Contemplative Man's Recreation," Walton mentions Dr. Wharton as "one who loves me and my art," presumably meaning that the doctor was an enthusiastic fisherman. The reason for the allusion was that Wharton described his dissection of a strange fish "almost a yard broad and twice the length." Perhaps fishermen's exaggerations were common even in those days.

During the birth of a child, the body of the baby—and particularly its head—has to navigate the narrow tunnel of the mother's pelvis. In part due to the human adoption of the upright posture, the space available for the passage of the infant is minimal, even in a woman with a normal pelvis. This is presumably a matter of natural selection, for if, as the result of evolution, the pelvic canal became any smaller, humanity would suffer rapid extinction —a fate that is liable to befall some small dogs who have been artificially bred to such an abnormal shape and size that the birth of a litter is almost impossible without a Caesarian operation. When this minimum space is further reduced by some abnormality of size or shape, obstetricians become concerned during pregnancy with measurement and the possibility of an obstructed labour.

In humans, the Caesarian section, now often a necessity in cases of obstructed labour, is the removal of the child through an incision in the abdominal wall rather than through the birth canal. It is an operation with a long history. Julius Caesar is said to have been born this way, hence the name of the operation which was initially performed only in case of the death of the pregnant woman. The subsequent history of the operation is gory and rumour-ridden, until the first known case in which both mother and child lived. That operation was performed in 1836 in Birmingham, England.

It is difficult to appreciate the hot disputes which occupied eighteenth- and nineteenth-century obstetricians about the posture and pathway of the child's head as it passed down the pelvic birth canal. The origins of pelvimetry began with the Scotsman William Smellie, the master of British midwifery in the eighteenth century. But the actual techniques and particular dimensions relevant to difficulties in childbirth were worked out by a series of anatomists and women's specialists over more than a century.

Jean-Louis Baudelocque, born in Picardy, France, in 1746, was the son of a surgeon and soon made a name for himself in Paris, taking up the familiar triad of anatomy, surgery, and obstetrics. He became a professor at the Charité Hospital at an astonishingly young age and became the acknowledged master *accoucher,*

called by nobility and royalty to deliver their wives.

Baudelocque had a strong interest in pelvic mensuration and advocated Caesarian section when the birth canal was obviously inadequate because of bony abnormalities of size or shape. This led him into grave dispute with some colleagues and he was actually brought to trial by a rival doctor when a patient died after a Caesarian. His book *L'Art des Accouchments,* published in 1781, in which he explained his highly detailed methods for estimating various pelvic dimensions, became the source of the standard technique of the time. One of the important measurements he made has not stood the test of time, but it took the introduction of X-rays to displace it. This was the external conjugate diameter, which was soon called Baudelocque's diameter or even more simply B.D.

Baudelocque believed that this external measurement reflected the true internal diameter of the pelvis from front to back, and for one hundred and fifty years great importance was given to this measurement as an index of the capacity of the birth canal. Although he died in 1810, two of his nephews carried on the Baudelocque name in French midwifery until two-thirds of the way through the nineteenth century.

Employing external measurements of the pelvis in the belief that they accurately reflect the internal diameters continued for a long time. One man who clearly pointed out the deficiencies of external pelvimetry has paradoxically been remembered by having his name linked with one of the external measurements which he himself decried.

Gustav Adolf Michaelis, a German doctor, wrote a book *Das Enge Becken* (The Narrow Pelvis), which was published in 1851, three years

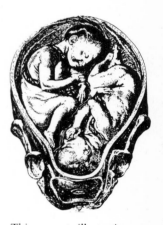

This accurate illustration of a normal presentation of twins in the uterus was published in 1754 in William Smellie's book *A Sett of Anatomical Tables.* Smellie, a famous London obstetrician, wrote a detailed account of the birth process and refined the use of forceps in delivery. He was satirized by Laurence Sterne in his novel *Tristram Shandy.*

This sixteenth-century depiction of the birth of Julius Caesar illustrates the popular belief that he was born by Caesarian section. Some claim, however, that the term Caesarian comes instead from the Latin *caedere*, to cut.

after his death. It has been said that no obstetrician could possibly understand the contracted pelvis unless he has read Michaelis's book. Yet a small and relatively unimportant section of the work was named after the author—the Michaelis rhomboid. This is a diamond-shaped area on the back of the female body, between the inner edge of the buttocks, the prominence of the hips, and the coccyx. Michaelis was intrigued by the different dimensions of this area in normal women and those with such pelvic abnormalities as rickets. His most important contributions, however, were in his critical attitude to the reliability of external pelvimetry, so it is ironic that eponymy played a trick upon him.

Gustav Michaelis, the son of a doctor, was born in Marburg, Germany, and graduated from the University of Kiel. He then went to Paris, the Mecca of doctors of that time. Although Michaelis never became a full professor at Kiel, he was active in medical and community affairs, but he developed a depressive illness and eventually committed suicide in 1848 by leaping under a railway engine. As well as medicine, Michaelis maintained a keen interest in natural history, the arts, and archaeology. His work in marine biology earned him another eponym, in that his observations on a phosphorescent organism caused it to be known as *perinidum Michaelis*.

Among the dozen or so obstetricians and anatomists who devoted themselves to pelvimetry in the eighteenth and early nineteenth centuries, only a few have had their names perpetuated. The name of Carl Gustav Carus has been linked to the curve formed by the axis of the birth canal as it sweeps first downwards into the upper inlet and then forwards, following the bend of the sacrum and coccyx. This arc was sometimes also called the axis of parturition.

The radius and conformation of this parturient axis has been the subject of intense research in obstetrics. In 1820, Carl Carus published a simple yet elegant essay on pelvic geometry that soon caused his name to be applied to the curve he described. The curve of Carus was actually not a true representation of the track of the foetal head as it passes through the birth canal. Soon after the publication of his volume *Textbook of Gynaecology*, it was criticized and confounded by Franz Carl Nägele, the German obstetrician, who put forward his own corrected version of the parturient axis. For a long time, however, most obstetricians preferred the simpler concept of Carus and many later authors adhered to his views.

Carl Gustav Carus was born in 1789 in Leipzig, a city that was the birthplace of so many brilliant medical men. He graduated in medicine from the famous university. When the new medical institute opened in Dresden, the twenty-five-year-old Carus moved there as professor of obstetrics. He later reached high office in governmental administrative medicine, but still found time for great literary output. Like so many doctors of past centuries, Carus had wide scientific interests especially in the field of natural history. He was, for example, an authority on insect anatomy. His artistic interests were equally remarkable and he was a friend of Goethe, Jenny Lind, and Henrik Ibsen. Indeed, he was acknowledged as one of the leading romanticists of his century and again it seems odd that medicine remembers him, as it does Michaelis, for a pelvic dimension of dubious value.

Franz Carl Nägele, who criticized Carus, was attributed with an eponym having to do with one of the manoeuvres of the foetal head as it entered labour. It was said to be an asynclytism, or lateral flexion, of the baby's head. This is often referred to as Nägele's obliquity, for it was Nägele who emphasized in 1819 that this was a normal part of the movement of the foetal head.

Studies in pelvimetry are linked to deeper and more complex understanding of completely normal births, births in which the child develops normally, but the mother's pelvic structure is either too small, or otherwise deformed, to allow for a normal birth, and of births in which the child is in a position in the womb which precludes a normal birth.

Three obstetrical techniques are associated with and developed from the consideration of difficult births. One is the Caesarian section, which was almost always fatal until the consistent use of anaesthesia, asepsis, and antihaemorraghic agents. It was contemplated as a possibility, however, in two instances. Until relatively recently, the survival of the mother did not matter since she was already baptized and her immortal soul safe, but concern was fixed on the unborn infant. Either it could be baptized in utero by a fanciful baptismal syringe and left to its fate, or it could be extracted by a crude Caesarian section. The other instance in which a Caesarian section was considered was when the mother's pelvis was too deformed, by rickets or some other debility, to permit a normal birth. A frequent modern use of the Caesarian section occurs when the mother's pelvic ring is too small in relation to the baby. But it was not until the middle of the seventeenth century that it was more or less established, in a book of uncertain authorship entitled *The English Midwife, Enlarged, Containing Directions to Midwives; . . . The Whole Fitted for the Meanest Capacities,* that contrary to earlier and quite baseless notions, the pelvic ring does not open up, like a split hoop, to allow the passage of the baby.

Another obstetrical technique, rather more useful and less dangerous than the Caesarian section, has been used at least since the second century A.D., when it was introduced by Soranus of Ephesus. This is podalic version, a procedure of reaching into the womb and turning and extracting a child whose position in the womb makes a normal birth impossible. Podalic simply means "by the feet." The midwife or doctor reaches in and grasps the child by the shoulder or whatever is accessible, turns it until the child's feet can be grasped and extracts it that way.

Podalic version, like many other medical techniques, discoveries, and insights available to the ancient and classical world, went into eclipse for about thirteen hundred years, at least among the medical establishment, until it was reintroduced by Ambrose Paré, a French physician who worked in Paris at the Hôtel Dieu in the early sixteenth century. Paré began as a rustic barber's apprentice, then trained as a dresser and "apothecary boy" in the unspeakable conditions of the Hôtel Dieu—a combination of poorhouse, insane asylum, hospital, and lying-in ward. He was known as a compassionate and gentle man. He did not write extensively about podalic version, but taught it to his assistants, one of whom saved the life of Paré's daughter by its application.

The third technique involved the invention and use of an instrument, the obstetrical forceps. Like so many medical techniques it was grossly abused, especially in its early history, and has had much notoriety as well as fame. The forceps consists of two wide flat blades curved to fit gently over the child's head. The blades are inserted and brought into position separately, then locked together firmly gripping the head. By turning and pulling gently, the child is extracted; any opening through which the head can go, the rest of the child can follow, since the diameter of the baby's head is larger than any other part of its anatomy. The forceps is used in two cases—either when labour is advanced and the baby's head gets wedged in the pelvic ring for too long a time, or when, because of the condition either of the mother

or child, it is advisable to terminate labour rapidly. Obviously, assessment of both these situations is a matter of discretion. It would seem that neither the inventor of the forceps, and members of the family obstetrical dynasty he founded, nor many subsequent practitioners, were very prudent in their use. Today, the tendency is to use forceps only in case of the direst emergencies, never for the mere convenience of the physician.

The forceps were invented in the late sixteenth century by the sons of one Chambellan, a Huguenot physician who fled from France to England to avoid the massacre of Huguenots on St. Bartholomew's night in 1571. Peter the Younger and Peter the Elder, by now Chamberlain or Chamberlen, became influential physicians and turned their attention to midwifery. The brothers kept their invention a secret, and on the basis of their claim that they could deliver patients when all others had failed, they attempted to control the instruction of midwives. The next generation, represented by a son of the younger Peter, still kept the secret and wrote a self-aggrandizing pamphlet adver-

The illustrations below, from a thirteenth-century English manuscript, are meant to indicate possible presentations of the child in the womb, including, in the lower right-hand corner, a breech presentation.

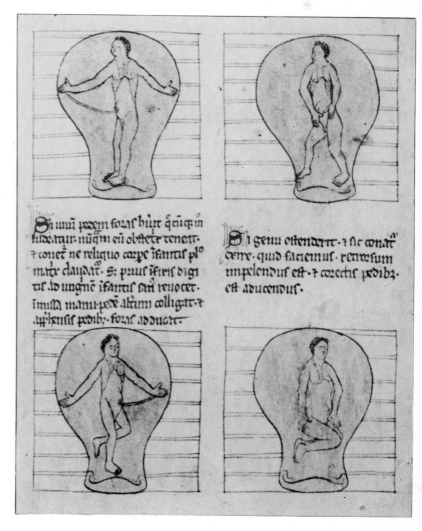

tising his skill. The most notorious Chamberlen was his grandson, Hugh, who said, in the middle of the seventeenth century, "My father, brother and myself (though none else in Europe as I know) have by God's blessing and our own industry attained to and long practised a way of delivering women in this case without prejudice to them or their infants." Perhaps they did, but not without at least one well-known and particularly brutal abuse.

For political reasons, Hugh retired briefly to France during the reign of Louis XIV. There he met François Mauriceau, the leading obstetrician in France. Hugh boasted to Mauriceau that he could, by means of his secret, deliver the most difficult case in a few minutes. Mauriceau placed at his disposal a woman dwarfed and deformed by rickets, for whom all efforts at assistance had failed. Hugh worked for three hours before he finally admitted he was having difficulties. The woman died from injuries caused by his manipulations. "To complete the story," Mauriceau wrote, "it should be remembered that six months before the occurrence of these events, this physician had come to Paris from England and boasted that he possessed a secret method by means of which he could, even in the most difficult case, promptly effect the delivery of the child, and had told the king's physician-in-ordinary that he would sell the knowledge of this secret for the sum of 10,000 Thalers."

Hugh Chamberlen fled from France to England, and in 1699 was forced to flee again to Holland, where he had better luck. He sold his "secret" to the Medico-Pharmacological College of Amsterdam. The college had the sole privilege of licensing physicians in Holland. To each licensee it sold the "secret" for a large sum. This continued until the secret was purchased by a group of what would now probably be called concerned citizens, and was made public. At this point they discovered

In this woodcut by Hans Holbein called *Death and the Priest,* Death leads the way for the priest who carries the sacrament to a dying person. The real version of this macabre procession terrorized patients in the lying-in ward where Semmelweiss fought to end puerperal fever.

that Chamberlen, or perhaps the college, had perpetrated a giant swindle. The device they had bought was a half, an entirely useless half, of the forceps. Hugh's son, Hugh Junior, became a prominent physician in England. At the end of his life he allowed the "secret" to leak out, and the instrument soon came into general use. Credit for the invention of the forceps is not always given to the Chamberlens, however, but to a Belgian, Jean Palfyne, who developed an instrument which he presented freely to the Paris Academy in 1721.

One additional important medical development belongs essentially in the realm of normal births. As the responsibility for assisting in childbirth passed from the midwife to the physician, there was a corresponding proliferation of maternity hospitals and lying-in wards in general hospitals. Early hospitals were pestilential places. The most primitive precepts of cleanliness, let alone asepsis, were unknown, as was any concept of the process of contagion. As a matter of course, doctors went insouciantly from dissecting diseased cadavers straight to perfectly healthy women in childbed. Consequently, Europe was ravaged by epidemics of childbed, or puerperal, fever, which is simply wound infection, blood poisoning caused by the contamination of the raw surface left in the uterus after childbirth. Between 1652 and 1862

This sixteenth-century woodcut shows a room in the pestilential Hôtel Dieu, the infamous lying-in ward, madhouse, and general hospital in Paris. To be confined there was considered a death sentence. The beds were each intended for two patients, but often five or six, regardless of sex or ailment, were crowded into each one.

there were two hundred epidemics of the disease. In 1773, an epidemic decimated the lying-in hospitals of Europe and after raging for three years reached a climax in Lombardy, where it is said that for one year not one woman lived after bearing a child.

One lonely medical hero fought, for a long time hopelessly, to end this plague. Ludwig Semmelweiss, born in Budapest in 1818, studied and practised in Vienna in the particularly miasmic lying-in hospital there. He became aware that in the section of the hospital where the poorest women were attended only by midwives (who never handled cadavers, diseased or otherwise) the death rate from puerperal fever was half that of the other section of the hospital where physicians were in attendance. Women being admitted to the hospital were aware of the same situation, for they went to great lengths to gain admission to the section served by midwives.

Semmelweiss was in charge of the section where the physicians practised. What struck him most forcibly, however, was the strange death of a colleague, a Dr. Kolletschka, who, while performing a post-mortem examination, received a puncture wound in the finger when the knife of one of his pupils slipped. He died in agony, his symptoms the same as those of puerperal fever. Semmelweiss said, "In the excited condition in which I then was, it rushed into my mind with irresistible clearness that the disease from which Kolletschka had died was identical with that from which I had seen so many hundreds of lying-in women die. . . . Day and night the vision of Kolletschka's malady haunted me. . . . In the case of Kolletschka the cause of the disease was cadaveric material carried into the vascular system; I must, therefore, put this question to myself: Did, then, the individuals I had seen die from an identical disease also have cadaveric matter carried into the vascular system? Yes."

When he decided this, Semmelweiss at once made each student wash his hands in chloride of lime, a crude antiseptic, before making examinations. Within seven months the death rate in Semmelweiss's section was one-tenth of what it had been. His discoveries were not recognized during his lifetime. He died in 1865 at the age of forty-seven, embittered and a victim of the same infection, from a small injury he sustained during one of his last operations.

The understanding and minimizing of the major risks of pregnancy and childbirth bring us full circle—to the new human being, the only creature who questions, restlessly, perpetually. The pressure of this questioning, more often than not done in the uncongenial climate of superstition, religious dogma, and, most

Ambrose Paré was not an educated man. He wrote in the vernacular and was ridiculed for his ignorance. He began his career as a barber's apprentice, then became an army surgeon, and was eventually acclaimed as the greatest surgeon of his day and served as physician to three successive kings of France. Paré, who was born in 1510, studied the works of Vesalius, Fabricius, and their contemporaries. He invented instruments and techniques to improve practical surgery and he tried to debunk such cherished superstitions as the use of healing amulets. One of his most important innovations was his treatment of battle injuries. He could not bear to pour boiling oil on gunshot wounds, as did his colleagues, and saved many more lives by simply applying dressings. Paré, who died in 1590, never claimed glory for himself: "I treated him, but God cured him."

recently, mechanistic rationalism, forced into existence the vast and intricately ordered body of knowledge we now have about the structure and function of the human body. The climate of inquiry cannot even now be described as free. We are as bound as were our ancestors by our preconceptions, by the limitations of our technology, by assumptions made on the basis of incomplete information, and by those political controls that determine which sort of research shall be permitted, or funded, and which shall not. Now, as then, there are fashions, styles, in thought as in dress. There is no way of knowing how much of what we "know" today will be modified, extended, revised, or even completely overturned tomorrow. Questions breed more questions, information more information.

The explorations into anatomy and physiology charted here belong to history, but the story of the discovery of the human body is by no means finished. More and more complex technology makes for geometric progressions, quantum jumps in the rate of change. From among the babies being born today, helped into life by our accumulated understanding of the human body, new explorers will come, to whom we will seem perhaps as crude, random, adventurous, and heroic as our predecessors do to us. Healthier, longer-lived, possibly with artificially accelerated thought processes, their natural senses amplified by a spectacular technology of which they will, with luck, remain in control, these new explorers will write the next interim report. There will never be a final report, so long as the creature keeps asking questions, and, what is equally important, requiring satisfactory answers.

The baptismal syringe was designed to baptize an unborn child in the womb. Whether it lived or died was of no consequence, so long as safety of its immortal soul was assured. For a long time such preoccupations obviated much concern with the life of the pregnant woman.

Corpus Delicti—A History of Human Dissection

Anatomy is a totally descriptive subject, thus its quality can only be as good as its basic material. Without human bodies for study by dissection, there could be no meaningful anatomy. Although such men as Aristotle and Galen were undoubtedly geniuses, their attempts to transpose aspects of animal anatomy to the human body sometimes led to ludicrous conclusions.

The first human dissections are mentioned only obliquely, but there must have been some in ancient Greece. It is known that in about 380 B.C. Diocles of Carystus dissected human foetuses, and the title of his now lost textbook was the first to bear the word anatomy.

Attitudes about tampering with the dead had been changing in Greece. The primitive fear of calling down the wrath of the gods on any desecrator declined by the time of the philosopher Heraclitus of Ephesus in the fifth century B.C. He was the Father of Metaphysics and among his rather dark beliefs was an emphasis on the clear separateness of body and soul. The unimportance of the mortal remains after death was also taught by Socrates, who lived from 470 to 399 B.C. His pupil Plato expounded on the ephemeral nature of the body, which was but a temporary shelter for the soul—a philosophy expressed, for example, in his *Phaedo*.

The Hippocratic school reaped some benefit from these changing attitudes and their writings, which date from about 400 B.C., on the suture lines of the skull and on details of the anatomy of the shoulder, indicate that some examination of the human body, however limited in extent, must have taken place at that time.

Aristotle, a little later in Greek history, seems, however, to have confined himself to animal dissection, accounting for some of his extraordinary errors of anatomy. In mainland Greece, as opposed to Hellenistic Egypt, human dissection appears to have been a little-practised procedure, in spite of the apparently liberal attitudes about it. Actually, it is questionable how broadminded the Greeks were on this matter. All through history, reflected even in relatively recent legislation, there has been a general reluctance, repugnance, or even abhorrence, to disturb the dead. Indeed, even today, controversy arising from social attitudes towards the treatment of the dead body surrounds the supply of cadaver organs for transplantation.

The body in this drawing, from a fourteenth-century French manuscript, is apparently suspended. The dissector is likely to have been some lowly person, perhaps only a barber, since until the sixteenth century, physicians did not deign to touch a corpse.

After these rather shadowy beginnings in Greece, the much bolder situation in Alexandria stands out in sharp contrast. The two great figures of the Ptolemaic medical school in Egypt around 300 B.C. were Herophilus of Chalcedon and Erasistratus of Chios. They did indeed dissect the human body and this was when anatomy began to be built on facts. There is ample evidence of extensive human anatomizing under their direction, on a scale that was to give rise to some scandal. The censorious figures of later and more restrictive times viewed the activities in Alexandria with horror, although the great Galen had no such retrospective condemnation to offer, merely stating that Herophilus was "the first to dissect both animal and human bodies." This must have referred to public dissection, because some limited anatomizing had been carried out in early Greece.

Both Herophilus and Erasistratus were later accused of human vivisection, of dissecting criminals while still alive. As with dissection of the dead, vivisection has been an evocative and emotive topic through the ages, currently—and perhaps with more justification—being of considerable concern in relation to experiments on animals.

The accusations came from three men, much later in time than the period during which the Alexandrians worked. The first accuser was Aurelius Cornelius Celsus, the aristocratic Roman who, although he was not himself a doctor, wrote the massive encyclopaedia which contained much excellent medical material. Writing in about A.D. 30, he twice referred to human vivisection at the Ptolemaic medical school. Although Celsus was in favour of using the human body to increase anatomical knowledge—a practice which by his time was coming to an end in the Roman Empire—he was opposed to vivisection.

More virulent in his condemnation was Quintus Tertullianus. Known as Tertullian, he was the earliest of the major Christian writers who lived in Carthage from about A.D. 155 to 222. He accused Herophilus of being a butcher who had dissected six hundred persons, with the implication that some of them were alive. He also condemned the opening of the human womb and accused Herophilus of killing foetuses as well as adults. Unlike the pre-Christian

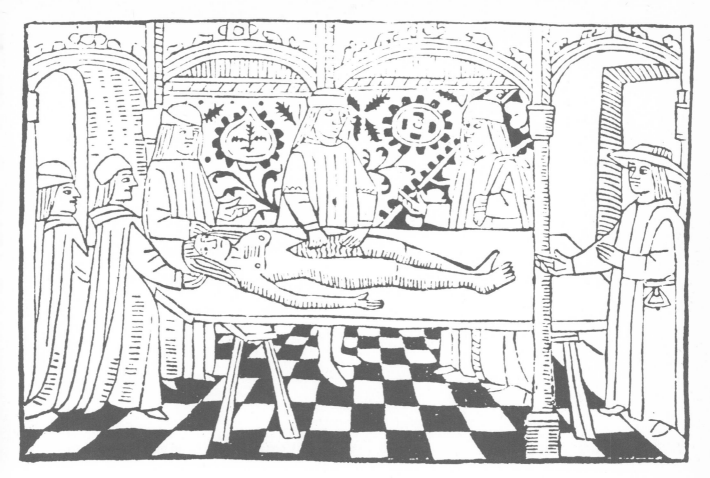

Celsus, Tertullian seemed motivated mainly by violent anti-paganism, and Herophilus was but one of many targets.

The third accuser of the Alexandrian anatomists was Saint Augustine, the bishop who lived from 354 to 430. He presumably repeated the condemnations of his Christian predecessor Tertullian. By that time the two Alexandrians had been dead for seven centuries.

As the medical historian Charles Singer pointed out, however, it is highly unlikely that there was human vivisection at Alexandria, for the simple reason that Galen—whose dislike for Erasistratus was remarkable—would never have missed such a good opportunity to pillory him if there had been any substance in the allegations. Galen devoted two books to abusing Erasistratus, but nowhere is there any mention of living victims.

When the great Alexandrian school declined after the era of Herophilus and Erasistratus, it seems that dissection petered out and that by 150 B.C. it had ceased altogether. Egypt was absorbed into the Roman Empire in 30 B.C. and from that time on, as far as practical human anatomy was concerned, there was no progress for more than one thousand years.

Galen had virtually no experience of the human body. His voluminous writings were derived from dissections of animals. He did his best to approach as closely as possible the human body by using monkeys and apes for much of his work, but the numerous errors and misconceptions that resulted are only too evident in his writings.

It seems that he saw at least one human body, this being, appropriately, at Alexandria where he studied for a time. It was probably a skeleton, although it may have had some dried tissues attached. One of Galen's accounts refers to seeing two skeletons, the first cleansed of its flesh by birds and the other by immersion in a river. He also mentioned, rather wistfully, that the sight worth seeing in the medical school at Alexandria was a human skeleton. It was from this skeleton that he wrote his book *Bones for Beginners*.

From Galen's time—which ended in A.D. 199—until the later part of the first millennium, anatomizing in the practical sense came to a complete halt. The dead body was still regarded with superstitious awe. It was believed that the spirit hovered near its dead body and, because there was belief in life after death, there existed uncertainty about resurrection of the body. Indeed, in the Talmudic writings of the ancient Hebrews there is a theory of a resurrection bone—the nucleus around which the body was

In this fifteenth-century woodcut of a dissection, the corpse seems to be a woman, a rare sight indeed on a dissecting table. Corpses were in any case not easy to come by, and were usually executed criminals, among whom were never many women. Another indication that the corpse might be a woman is that the examiners seem to be looking into the abdominal and genital area of the corpse; female sexual organs were a subject of controversy for centuries.

to be reassembled. And although St. Paul said, "It is sown a natural body, it is raised a spiritual body," early Christians nevertheless feared the consequences of tampering with the dead natural body.

Intellectuals in the Islamic world preserved the ancient writings for Western posterity, but there were no advances made in anatomy, for the Moslem religion, too, forbade dissection. Only a few whispers have come down through history of any stray opportunities that there might have been to learn more about the human body. Theophanes, for example, mentions in a book written in about A.D. 810 that in the previous century a Christian renegade had been turned over to surgeons in Alexandria, who "opened the still-living victim from pubis to chest to study the human structure."

In the ninth century, Johannes Mesaweight, a Jewish doctor working in the Muslim world, bemoaned the prohibition of dissection on religious grounds and regretted that instead he was forced to work on monkeys. He further observed that if it had not been forbidden, he would have been "willing to give my mentally retarded son for vivisection to see if the cause of his stupidity could be discovered."

The beginnings of dissection in the universities of southern Europe are clouded with uncertainties and errors of documentation, but movements in that direction began in the twelfth and thirteenth centuries. Frederick II, Emperor of the Holy Roman Empire, was one of the first to be involved. It is said that in 1238 he ordered that a public dissection should take place every fifth year at the medical school in Salerno. He also decreed that no surgeon be allowed to practise unless he studied anatomy for one year. These rules were shadowy, however, and may not have been carried out. The evidence, too, is conflicting, for the physicians of Salerno, at that time the premier medical school of Europe, declared dissection of the human to be "a horrible action, especially to us Catholics."

The cradle of modern dissecting was undoubtedly at the University of Bologna in northern Italy, where a faculty of medicine existed as early as 1156. The university had by far the best-known law school in the then-known world and it was as a by-product of legal necessity that organized anatomical dissection came into being. The lawyers controlled the medical faculty during the early years and it was not until 1306 that the medical school at Bologna was allowed to choose its own head.

The first post-mortems, probably in the middle of the thirteenth century, were performed at Bologna as medico-legal procedures, to investigate deaths which had some forensic significance. The first known one was recorded by William of Saliceto, a surgeon at Bologna and a teacher on the medical faculty there. He lived from about 1215 to 1280 and in his book *Surgery and Anatomy* he mentioned a case examined in about 1275. The post-mortem was carried out to discover whether certain wounds of the chest had been severe enough to cause death. Other similar medico-legal autopsies were mentioned in 1295 and 1300. There must have been many more which were recorded in documents that have not survived.

None of these early autopsies seem to have been carried out for anatomical purposes. In 1302, however, a man named Azzolino died under suspicious circumstances in Bologna and was thought to have been poisoned. After a detailed examination by two physicians and three surgeons no evidence of any poison was forthcoming, but the surviving autopsy report significantly mentions that the result was arrived at "from the evidence of our senses and of our anatomization of the parts."

At about the same time, a Bolognese surgeon writing about the breast changes in pregnancy, commented that "I have not seen the dissection of the breast of a pregnant woman," the clear implication being that he had seen other dissections.

From this time onwards, dissection for anatomical purposes, as opposed to autopsy investigations, undoubtedly took place with increasing frequency, probably limited mainly by the scarcity of bodies. Men like Thaddeus of Florence, Bartolomeo da Varignana (who was the leader of the team who performed the autopsy on Azzolino), Henri de Mondeville, and Mondino de Luzzi, one of the first medieval writers on anatomy, were all involved in dissections in the first years of the fourteenth century.

Henri de Mondeville was a Norman who studied and taught at the University of Bologna. In 1304 he returned to teach at the University of Montpellier in France. While he was at Bologna he did some diagrams of dissections which he took back to his own country. His illustrations were concerned, however, with methods and techniques of dissection, rather than the study of the anatomy of the body.

From Mondino's textbook *Anathomia,* published in 1316, an autopsy manual of forty small pages, we learn that a dissection was a lengthy affair, sometimes a marathon. The relatively infrequent windfall of a body had to be taken advantage of quickly, mainly because of the rapidity of putrefaction. Dissection went on almost continuously for perhaps four days and even nights. The most perishable parts were examined first. The usual order was the abdominal cavity, then the chest, followed by the head, and last the limbs.

Mondino referred to a stern prohibition of

dissection issued by Pope Boniface VIII in a Papal Bull dated 1299. This was not intended to repress anatomical dissections, but to stop the boiling of the bodies of dead Crusaders. The reason for this curious practice was that when Crusaders from northern countries died in the Near East (much more often from disease than crusading) it was a practical impossibility to ship their bodies back to their homelands for burial. Often, as the next best thing, their comrades collected the skeletons by boiling the flesh from the bones, it then being a much easier task to send home the dried bones. Apparently Boniface forbade this, but it was still frequently done. Mondino observed that the bones of the skull were better demonstrated if they were first boiled, although he added, "Owing to the sin involved in this, I am accustomed to pass them by."

The practice of human dissection must have spread quickly across Europe, no doubt because of the close community of academics which existed in those days much as it does today, although then the lines of communica-

tion were more personal. Within a few years either way of 1300, dissections were being performed in England. An illustration now in Oxford's Bodleian Library depicts a surgeon eviscerating a body and showing the organs to a physician and a monk standing nearby. The body was female, which was unusual, as most dissection subjects at this time were male criminals from the gallows. Indeed, it was customary in some places for the anatomist to specify the actual method of execution, to preserve certain parts from damage. During some periods, it was mandatory, rather than permissive, for the executed victim to be anatomized, this being an added part of the punishment to be anticipated by the victim, rather than a gesture towards the anatomists.

Although bodies for dissection were obtainable by legitimate means, particularly by the official disposal of executed criminals, the increasing need of medical students and doctors had to be supplemented by more clandestine methods. It was common for students to go off on nocturnal body-snatching expeditions with

This 1610 engraving of the anatomy school at Leyden illustrates, more than anything else, the persistence of a slightly macabre medical school humour, including the preserved cadavers (or perhaps posing students) mounted in the centre, both with hats, and one with a pipe. The theatre seems to be a place of popular resort, complete with a casual family scene in the lower right-hand corner.

L'ANATOMIE DE LEYDE.

the full approval—indeed the active urging and encouragement—of their teachers. Sometimes they were caught by the authorities. In Bologna there is a record of the prosecution in 1319 of four medical students for grave-robbing. Some embarrassment must have been caused to the medical faculty in succeeding years, for a rule of the university dated 1405 decreed that no one might acquire a body without the permission of the Rector.

There were quite elaborate regulations designed to spread experience of anatomical dissections as widely as possible. No more than twenty students were allowed to watch the dissection of a male body, but if it was a woman, thirty were permitted, because the rarity of such an opportunity warranted the extraction of the maximum teaching value. The students had to be in their third year of study and if they had already seen a male dissection, they had to wait another year before attending again. Once a student had seen two dissections, he could not come again unless there happened to be a female body, which was a chance obviously too good to miss. The supply of cadavers was limited to executed criminals, and fewer women than men were sentenced to die for criminal acts. A pregnant female cadaver was even rarer. This is the reason that the study of female anatomy fell so far behind that of the male.

In 1442, Bologna decreed that the city authorities were to make two bodies available each year to the medical school, one male, the other female, unless they could only manage to get two males. The bodies had to be those of executed non-Bolognese, whom they called "foreigners."

By 1341, Padua followed Bologna in the practice of dissection and then Pavia, Florence, Siena, and other Italian schools took up the practice of public dissection in their medical faculties. All followed the same procedure, with the professor sitting high in his pulpit. He would read from the grossly inaccurate works of the ancients or the more recent re-writes of Galen, such as Mondino's *Anathomia,* while his junior academic, called the ostensor, would stand near the body and direct with his wand where the body was to be cut. The task was carried out by the actual dissector, who was either a menial servant or perhaps a barber or surgeon, both of whom rated quite low in the medical hierarchy of the time. Only rarely, as in the case of Mondino and later of Vesalius, would the professor deign to come down from his chair and touch the body himself.

During the next few centuries, the pattern remained much the same. After the censures and prohibitions of the earlier medieval period, public attitudes towards human dissection became more relaxed, particularly in Europe. By

The corpse is the least interesting aspect, although it remains the focus, of this strange seventeenth-century German engraving of an anatomy theatre. The gentleman in the lower right-hand corner is showing the lady, or perhaps even trying to sell her, what can only be a complete human skin. The two skeletons in the lower centre are miming Adam and Eve in the garden, complete with apple and serpent.

For the rest, the engraving is an emphatic memento mori, a reminder of death, exactly analogous to the skulls philosophers and scholars kept on their desks to remind them of their mortality. All around the room, the banners, in Latin, sported by the skeletons are reminders of mortality: "Birth is the beginning of those who are about to die," "We are dust and shadow," "Being born, we die," and, in the hands of a skeleton, what else could "Know thyself" mean. All the banners are pointed towards the corpse, as if to reinforce the message.

These two paintings of anatomy lessons, the one on the right Italian, the one below seventeenth-century Dutch, illustrate the varying states of preservation of cadavers. The one in the painting below seems to be fairly recent, while the one on the right, unless it had been flayed to demonstrate the muscle structure, a not uncommon practice, appears in a far more advanced state of decomposition.

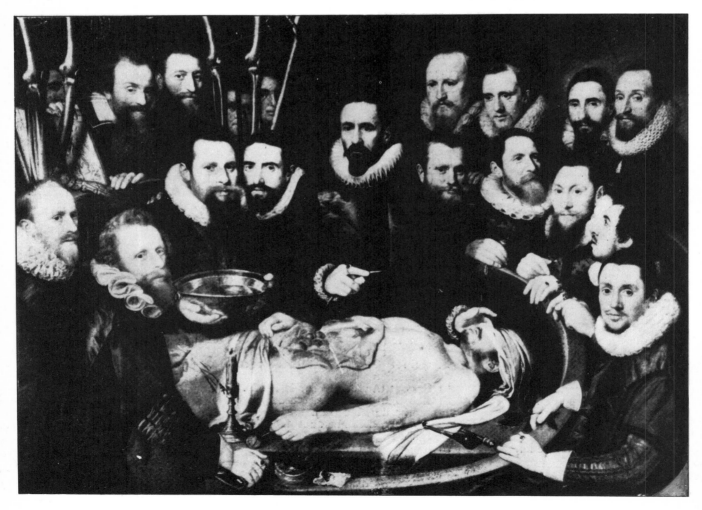

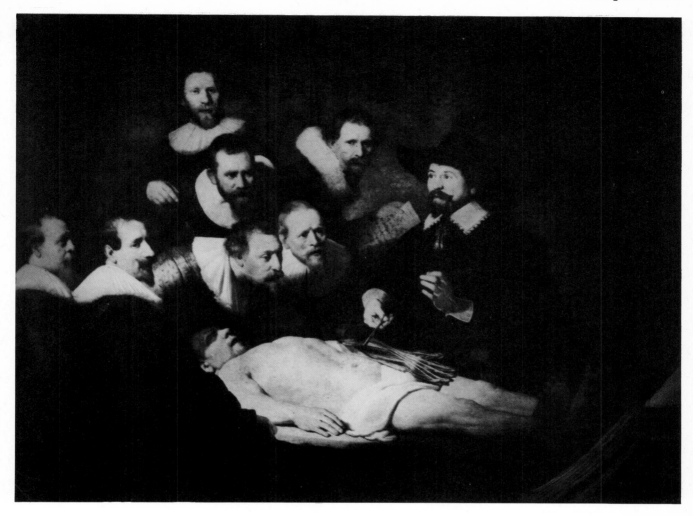

the time of the Renaissance at the end of the fifteenth century, many artists, including Dürer, Michelangelo, Donatello, Raphael, and Leonardo da Vinci, were dissecting bodies to improve their anatomical knowledge, an indication of the easier climate of public opinion.

Although it was as an artist that Leonardo began to study surface measurements, proportions, and anatomy of muscles, it was as an anatomist that he studied the brain, heart, digestive system, and foetus. Indeed, he devised one technique which is still used today; he injected wax into the body cavities to reveal their exact structure. He performed dissections on at least thirty bodies. When he was in the service of the Borgia family he performed his dissections by candlelight in the mortuary of Rome's Santo Spirito. From these dissections he made more than seven hundred and fifty accurate drawings and wrote one hundred and twenty anatomical notebooks.

In one of his notebooks, in justification of anatomical illustration, Leonardo wrote: "You who say that it is better to watch an anatomical demonstration than to see these drawings, you would be right if it were possible to observe all the details shown in such drawings in a single figure, in which with all your cleverness you will not see or acquire knowledge of more than some few veins, while in order to obtain a true and complete knowledge of these, I have dissected more than ten human bodies, destroying all the various members and removing the minutest particles of flesh which surrounded these veins, without causing any effusion of blood other than the imperceptible bleeding of the capillary veins. And as one single body did not suffice for so long a time, it was necessary to proceed by stages with so many bodies as would render my knowledge complete; this I repeated twice in order to discover the differences. And though you should have love for such things you may perhaps be deterred by natural repugnance, and if this does not prevent you, you may perhaps be deterred by fear of passing the night hours in the company of these corpses, quartered and flayed and horrible to behold; and if this does not deter you, then perhaps you may lack the skill in drawing, essential for such representation, and

Rembrandt's famous painting *Professor Tulp's Anatomy Lesson,* above, evoked the dramatic quality of public anatomies, and crystallized the relationship between science and art. The painting is largely focussed on the striking clarity and accuracy of the exposed arm and hand muscles, and the spectators are obviously gripped by the elegant demonstration.

ANDREAE VESALII
BRVXELLENSIS, SCHOLAE
medicorum Patauinæ professoris, de
Humani corporis fabrica
Libri septem.

CVM CAESAREAE
Maiest. Galliarum Regis, ac Senatus Veneti gra-
tia & priuilegio, ut in diplomatis eorundem continetur.

In the illustration on the title page of Vesalius's great work, left, the exigencies of anatomical demonstration during the Renaissance are made clear. Perhaps because the body is that of a woman, the demonstration area is terribly crowded. Vesalius himself is doing the dissection. He was one of the first anatomists not only to do his own dissections, but to lecture directly from the cadaver. His flaunting of scholarly medical tradition for the sake of empirical demonstration caused much scandal.

if you had the skill in drawing, it may not be combined with a knowledge of perspective; and if it is so combined you may not understand the method of geometrical demonstration and the methods of estimating the force and strength of muscles; or perhaps you may be wanting in patience so that you will not be diligent."

The opportunities of even the best known of the Renaissance anatomists actually to work on human bodies were, however, surprisingly limited. The physical availability of anatomical material remained a problem.

It was always easier to get permission for a post-mortem examination to determine the cause of death than for an anatomical dissection. Andreas Vesalius related how there was little difficulty in getting permission for an autopsy on the young daughter of a nobleman in Brussels; the cause of death was said to be a chest complaint caused by corsets which were too tight.

Getting dissection material, especially on female bodies, was another matter. All of Vesalius's experience with the female generative organs, for example, was derived from only six dissections, of which three had to be used for public demonstrations. Of the others, which he used in the preparation of his monumental work, the seven-volume *De Fabrica* of 1543, one was that of a six-year-old girl, stolen from her grave by one of his students and already half decomposed. Of the remaining two, one was a pregnant woman who had been

murdered (and he had to make a rapid autopsy on this for medico-legal purposes), and the other a woman who had hanged herself. The anatomy of the female organs in *De Fabrica* was largely based on dissections of this one unfortunate suicide.

When he was a student in Paris, Vesalius stole bodies and made many visits to the Cemetery of the Innocents to search for bones. He spent many long hours there turning over the piles of bones and "was gravely imperilled by the many savage dogs." He went to the cemetery with a friend, Mattheus Terminus, who later became a distinguished physician. They became so expert at recognizing the different bones of the human body, as they competed with the dogs in the cemetery, that they would win wagers with other students that even blindfolded they could identify, by touch alone, every bone that was handed to them during a half-hour period.

Vesalius studied medicine at Paris under Sylvius, the most popular medical teacher of his time and a devout Galenist. Sylvius also taught anatomy, but he knew Galen better than he knew the human body and frequently the ostensor could not find the structures Sylvius described as he recited them from Galen's writings. Vesalius and his fellow student Michael Servetus dissected bodies and presented Sylvius with evidence. This was considered an unforgivable offense. Nevertheless, Vesalius realized that he was an anatomist and that his professor was not. It was at this point that he be-

Shown below is the woodcut Vesalius included in the second book of *De Fabrica* showing his complete collection of dissecting instruments, some of them fairly horrific.

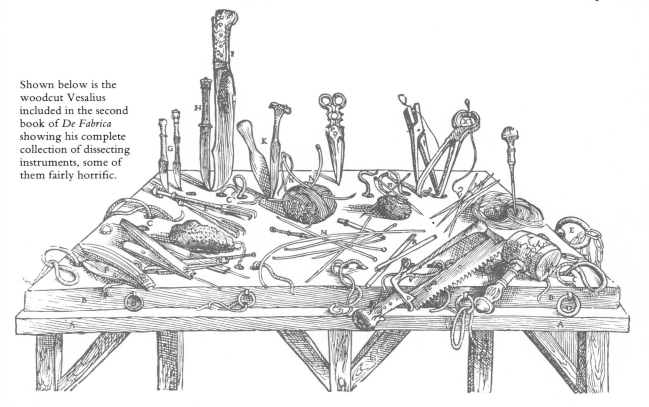

The illustration above is a reproduction of the title page from a book by Eustachius printed in 1722, showing Eustachius himself performing a dissection in an anatomy theatre of the time.

In the late sixteenth-century Dutch anatomy theatre left, the herbs are strewn on the floor to minimize the stench of decomposition.

came determined to devote himself to anatomy.

Vesalius gave an account of a typical student activity, the theft of a body from a gibbet. In the autumn of 1536, out walking with a friend (this time Gemma Frisius, who was later to become a famous mathematician), they came across a roadside gallows and found the body of a robber which had dried up after being bound to a tall stake. Vesalius climbed the pole and managed to pull off an arm and a leg, which in successive trips he smuggled home under his cloak. Anxious to acquire the trunk and head, he let himself be shut outside the city after dusk. He then tore down the rest of the body which was chained to the stake. Hiding this nearby, he managed during the next few days to smuggle it bit by bit through the city gate. In the elaborately illustrated chapter headings of his *De Fabrica,* Vesalius includes a drawing of a resurrection party, as these nocturnal body-snatching exploits were called.

In 1537, when Vesalius accepted the chair of surgery and anatomy at the University of Padua, he was the first man to receive a salary as professor of anatomy in any university. He was a wonderful teacher, acting as dissector himself and working with great technical skill. In *De Fabrica* he corrected more than two hundred errors in Galen's work and showed, once

and for all, what the structure of the human body really was. But after the publication of *De Fabrica* Vesalius had to contend with the ill will and ignorance of many of his colleagues and the threats of the Church. He burned his unpublished works and left Padua. His scientific career was over and he made no more original observations.

Meanwhile, at the end of the fifteenth century public anatomies began to be held in Europe. The earliest of these were simple, usually taking place in the town square and attended by passers-by. As time went on these anatomies became elaborate dramatic performances for paying audiences. Public dissections were planned months in advance and were advertised in handbills. Usually, the public anatomy was held once a year, at the beginning of winter. It had to take place in cold weather because of the perishability of the cadaver and the dissection would take three or four days to complete. The audiences paid large sums for admission and, less squeamish than people today, they were attracted by the drama and excitement of observing the intricate structure of the human body. By the seventeenth century, these public shows were held in amphitheatres specially constructed for the purpose.

The great expansion of population during the Industrial Revolution and the corresponding

increase in the number of medical schools caused the supply of bodies for dissection to be far outstripped by demand. This led to a scandalous state of affairs in Britain and the United States, although it seems that on the continent of Europe the crisis was less acute because of different official attitudes.

Since the time of Henry VIII, English law permitted only the bodies of executed murderers to be anatomized. By the end of the eighteenth century, however, this source was woefully inadequate. In 1747, William Hunter established a school of anatomy in London. By 1793 there were two hundred anatomy students in the city and by 1823 they increased to one thousand.

The stigma of murder had come to be attached to bodies for dissection and, consequently, public attitudes were antagonistic to anatomizing in Britain. And there was considerable resistance even to the idea of permitting any other legitimate source of supply. In Europe, more liberal codes and perhaps more autocratic governments allowed the bodies of poor or friendless people to be handed over to the medical schools. In some countries, including Germany, this was a fixed rule rather

than a permissive law and the supply was quite sufficient.

But in Britain—and in the United States until an even later period—the practice of body snatching became an industry. The trade of the "resurrection men" attracted the lowest dregs of a poor society, and criminal elements added grave robbery and murder to their basic occupation. It was said that in 1828 there were two hundred people in London engaged in the illegal procurement of bodies for medical schools. About eight hundred bodies a year were supplied and most of these were illegally exhumed from recent burials, a scandal which inflamed the public, especially after the notoriety of the Burke and Hare episode in Edinbugh, which ruined the reputation of Dr. Robert Knox, one of the anatomists. In London, the government was accused of turning a blind eye because of the need of universities, and armed guards stood watch on the graves of wealthier citizens. The price of a body rose from two pounds sterling to fourteen pounds (a vast sum in those days), and by 1830 many British students had to go to Paris to get anatomical experience.

The outrageous situation eventually forced

The pen and wash drawing, above, by Rowlandson, and the engraved scene in Surgeon's Hall, right, by Hogarth, both depict eighteenth-century dissecting rooms. The Rowlandson is a more or less straightforward representation, while the Hogarth, which is the last in a series of four plates called *The Reward of Cruelty*, is harsh caricature. The body is that of an executed murderer, still with the rope around his neck, and the heightened horror of the dissection proceedings is obviously meant to point a moral.

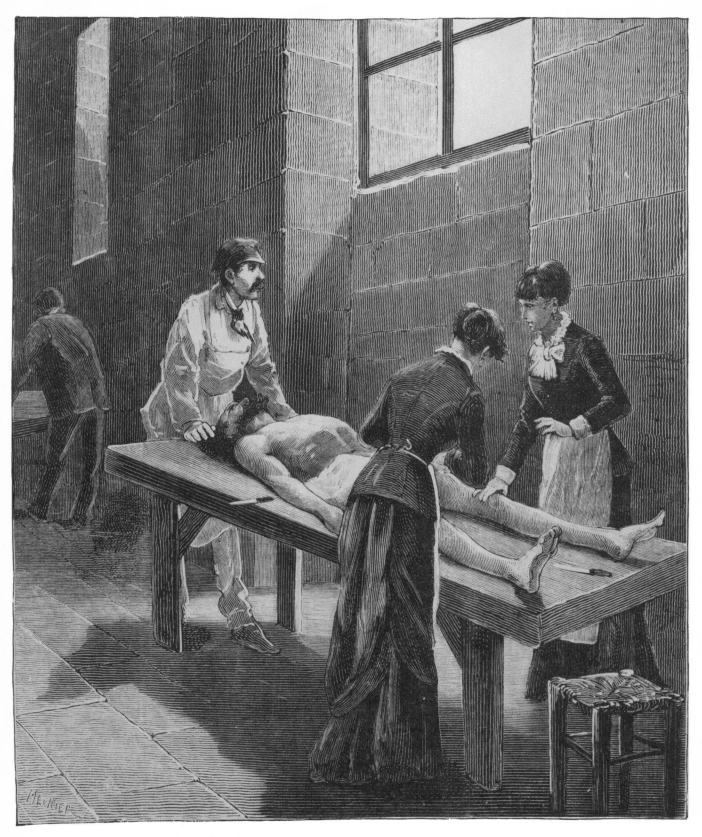

This unusual French engraving, dating from the
1880s, shows two young women medical students
beginning to dissect a cadaver at the Medical
Faculty in Paris.

government action, stimulated by an Anatomical Society which lobbied for legislation to provide a legitimate supply of cadavers. In 1828, a Select Committee of Parliament investigated the situation and in 1832 the first Anatomy Act was passed, undoubtedly helped in its passage through the House of Commons by the recent Burke and Hare scandal. Strong opposition delayed earlier attempts at passing such an Act, symptomatic of the reactionary feelings of a large part of the community. That Anatomy Act and its successors of later years regulate the supply of bodies to universities to this day, with a strict system of government inspectors and legal safeguards.

In the United States, the first legislation was even earlier than in Britain; Massachusetts passed an Act in 1831, a year before the British initiative. But each state has different laws and in some parts of the United States body snatching continued until the end of the nineteenth century, creating a situation even more chaotic than in England earlier in the period. Some states have mandatory, rather than permissive, laws. This means that bodies must be given up for dissection if they are unclaimed. This stricter attitude seems to reduce, rather than inflame, public opposition to dissection.

In the legal systems of all countries there is now a common provision that bodies must be treated decorously and buried decently at the expense of the medical institution, within a certain period. In Britain, for example, this is two years. Thus over a long road of two and a half thousand years' duration, the vicissitudes

The charming drawing at the left is of the artist Daumier, apparently carrying on the long and honourable tradition of medical students by snatching a quick lunch while examining a corpse.

of human anatomizing seem finally to have been brought to a sensible and humanitarian conclusion.

No longer does dissection hold risks for the dissector. In past centuries, it was not uncommon for anatomists to die from blood poisoning from infections they got as the result of handling poorly preserved bodies. Dissection was indeed dangerous until Louis Pasteur demonstrated the cause of infection and Joseph Lister began his campaign for antisepsis in 1869. When such antiseptics as carbolic acid or formalin solutions were injected into the blood vessels of cadavers the risk of infection was eliminated.

No longer are public dissections made amid scenes which resemble the marketplace, with dogs devouring the entrails. Today's dissecting room looks like an operating theatre, a clinical calm replete with closed-circuit television. No longer are charges of vivisection levelled at the academic staff and no longer do medical students have to creep out to rob gallows in the dead of night.

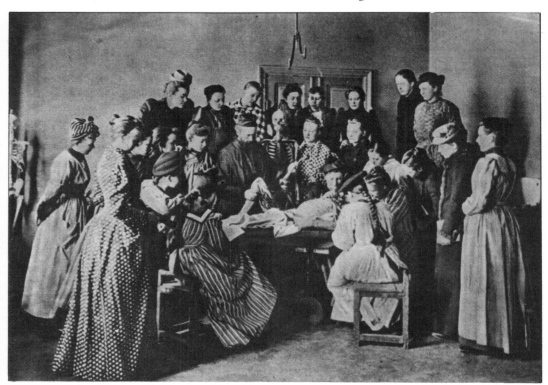

The photograph left, dating from 1880, is an unprecedented sight indeed, since a doctor is giving an anatomy lesson to women. The picture is Swedish and the women are probably trainee midwives.

Index

Page numbers in **bold face** type denote illustrations.

A

Acetabulum, **20**
  derivation 24
Acetylcholine, 95
Achilles tendon, **20**, **31**
Achillini, description of salivary duct, 59
Acoustic(s)
  and the ear, 100
  meatus, external, **96**
Adam's apple, **48**, **50**
Addison, Thomas, 132, **132**
  on adrenal gland, 131
  on white cells, 113
Addison's disease, 131
Adenographia (Wharton) (1656), 61, 160
Adenoid, **111**, 117
Adrenal cortex, **128**, 132
Adrenal gland, **128**, 131
Adrenaline, 95, 131, 132
Aelby, Christopher, 30
Aelby's muscle, **20**, 30
Aetius, **14**
d'Agrate, Ferrai, ovary, 144
Air
  and breathing experiments, 54-56
  passages, eponyms, 49
  pump, Robert Boyle, **53**
Albertus Magnus
  De Secretis Mulierum, 152
  embryology, 151-152
Albini, Giuseppe, 44
Albini's nodules, **34**, 44
Albinus, Bernard, muscles, 30-31
Albinus's muscle, **20**, 30
Albucasis, on thyroid gland, 130
Alcmaeon of Croton
  on auditory tubes, 96
  brain, 77
  early anatomy, 9
  on eyes, **102**
Alcock's canal, 31
Alexandria
  dissection of heart, 36
  early medical papyrus, 59
  early medicine, 12
Allen, Charles, *The Operator for the Teeth*
  (1685), 69
Allen, Harrison, 26
Allen's fossa, **20**, 26
Alveoli, **48**
Amnion, **150**, 152
Ampulla of Vater, **58**, 62
Anathomia (Mondino de Luzzi) (1316), 16,
  dissections, time factor, 168
  frontispiece (1493 ed.), **17**
  three-chambered heart (1513 ed.), **38**
Anatomical Notebooks (Leonardo da Vinci),
  22-23, **22**

Anatomical Society, legislation for dissection,
  181
Anatomy
  Anatomy of the Human Gravid Uterus (Hunter)
    (1755), 167
  Anatomy of Melancholy (Burton), 71
  Anatomy of Plants (Malpighi) (1670), 119:
    (Grew) (1682), 119
  early nomenclature, 8
  Galen's contribution, 13
  Guy de Chauliac (1363), **21**
  lesson, **174**, (175), **176**
  Padua, **16**
  paintings, **172**, **173**
  school, Leyden, **169**
    casual approach, **169**
  surface, artistic interest, 16-17
  terminology, 8, 38
    source, 21
  theatre, dissection, 176
    German engraving, **171**
Andernach's ossicles, **20**, 25
Aneurysma of Eysson, **150**, 160
Angle of Louis, 26
Animal spirit, 14, 80
Annulus of Zinn, **102**, 107
Anti-diuretic hormone, 130
Antrum of Highmore, **48**, 51
Anus, **58**
  eponyms, 63-64
Aorta, **34**
  ascending, 35
Apex, heart, **34**, 36
Appendix, **58**, 60
Aqueduct
  of Fallopius, 97
  of Sylvius, 77, 87
Aqueous humour, **102**, 104
Arab teachings, Dark Ages, 14
Aranzi, Giulio, 44
Aranzii, corpora, **34**, 44
Arataeus of Cappadocia, diabetes, 132
Areola, **140**, 149
Argive Hereum, statue showing pectineus
  muscle, 32
Aristotle, 11, **11**
  Arabic translation, 14
  on brain, 78
  early work in anatomy, 10-11
  embryology, 151-152
  genito-urinary system, **70**, 70-71
  on heart, 36
  Historia Animalium, 70
  on male organs, 134-135
  Masterpieces, 152
  on pituitary, 129
  on "psyche", 11
  on structure of teeth, 68
  on reproductive system, 141-142

Arnold, Julius, 113
Arnold body, 113
Arnold of Villanova
  innominate canal, 20, 25
  on urinoscopy, 71, 73
L'Art des Accouchments (Baudelocque), (1781),
  161
Arteries, 38
  "contain air", 36
  eponyms, 44-45
  of Zinn, **102**, 107
Aschoff, Ludwig, **34**, 47
Aselli, Gaspar, 61, **61**
  first colour in medical textbook, 60
  on lacteals, 60, 61
  De Lactibus Sive Lacteis Venis, 61
  on lymph, 116
Astrology in Dark Ages, 15
Atlas bone, **20**, 25
Atrioventricular bundle, 47
Atrium
  left and right, **34**, 38
Auditory canal, **96**, 98
Auerbach, Leopold, 63
Auerbach's plexus, **58**, 63
De Aure Humana (Valsalva) (1704), 98, **99**
Auricle, **96**
Avicenna, **14**, 15

B

Baer, Karl Ernst von, 123, **123**, 156
  cell division, 121
Baer's cavity, **150**, 157
Baer's membrane, **150**, 156
Baglivi, Giorgio, plain and striped muscle,
  28-29
Baillarger, J.G.F., 84
Baillarger, external and internal lines of, **77**,
  84
Balance, maintaining, 97
Balbiani, Eduard, 156
Balbiani's body, 156
Ball, Charles Bent, 63
Ball's valve, **58**, 63
Banting, Frederick, insulin, 133
Baptismal syringe, 162, **165**
Bartholin, Caspar, 142, **142**
Bartholin's gland, **140**, 142
Bartholin, Thomas, lymphatic system, 116
Bartolomeo da Varignana, 16
  dissections, 168
Basal ganglia, 86
Basilar membrane, **96**
Baudelocque, Jean-Louis, 161
Baudelocque's diameter, 161
Bauhin, Caspar, 61, 63
Bauhin's gland, **58**, 61
Bauhin's valve, **58**, 63
Bayliss, William, 130, **130**

Bibliography

Peter Edward Baldry, *The Battle Against Heart Disease.*
London: Cambridge University Press, 1971

O. Bettman, *A Pictorial History of Medicine.* Springfield:
Charles C. Thomas, 1956

Buchanan's Manual of Anatomy, 8th edition, edited by
F. W. Jones. London: Bailliere, 1949

Arturo Castiglione, *A History of Medicine.* New York:
Alfred A. Knopf, 1941

Emily Blair Chewning and Dana Levy, *Anatomy Illustrated.*
New York: Simon and Schuster, 1979

Edwin Clarke and Kenneth Dewhurst, *An Illustrated History
of Brain Function.* University of California Press, 1972

Logan Clendening, *Source Book of Medical History.*
New York: Dover Publications/Henry Schuman, 1960

Jessie Dobson, *Anatomical Eponyms,* 2nd edition. London:
Churchill-Livingston, 1962

Sir Michael Foster, *Lectures on the History of Physiology.*
Cambridge University Press, 1901

Kenneth Franklin, *A Short History of Physiology,* 2nd edition.
London: Staples Press, 1949

Fielding Hudson Garrison, *An Introduction to the History of
Medicine.* London: W. B. Saunders Co., 1929

Henry Gray, *Anatomy, Descriptive and Surgical,* 1901 edition,
edited by T. P. Pick and Robert Howden. Philadelphia:
Running Press, 1974

Gray's Anatomy, 35th edition, edited by Roger Warwick and
Peter L. Williams. London: Longman Group Ltd., 1973

Douglas Guthrie, *A History of Medicine.* London: T. Nelson
& Sons, 1945

Earle Hackett, *Blood, the Paramount Humour.* London:

Jonathan Cape, 1973

Howard W. Haggard, *Devils, Drugs, and Doctors.* London:
William Heinemann Ltd., 1929

Roger Lewin, *Hormones: Chemical Communicators.* London:
Geoffrey Chapman, 1972

Roberto Margotta, *An Illustrated History of Medicine.*
London: Paul Hamlyn, 1968

Jonathan Miller, *The Body in Question.* London: Jonathan
Cape, 1978

Noel Joseph Needham, *A History of Embryology,* 2nd edition.
Cambridge University Press, 1959

Charles O'Malley, *Andreas Vesalius of Brussels.* University of
California Press, 1964

Carl Sagan, *Broca's Brain.* New York: Random House, 1974

J. B. de C. M. Saunders and Charles O'Malley, *The
Illustrations from the Works of Andreas Vesalius.* New York:
Dover Publications, 1973

Science, Medicine, and History, Vol I, edited by Edgar
Ashworth Underwood. London: Oxford University
Press, 1953

Charles Singer, *The Evolution of Anatomy.* London: Kegan
Paul, Trench, Tubner & Co., Ltd., 1925

Henry Alan Skinner, *The Origin of Medical Terms,*
2nd   edition. Baltimore: Williams and Wilkins, 1961

Leonard A. Stevens, *Explorers of the Brain.* London: Angus
and Robertson, 1973

Harry Wain, *The Story Behind the Word.* Springfield: Charles
C. Thomas, 1958

Kenneth Walker, *The Story of Medicine.* London: Arrow
Books, 1954

Acknowledgements

The publishers wish to thank the photographers and
agencies listed below for their help in providing material for
this book. The following abbreviations have been used:
tl top left; *ct* centre top; *tr* top right; *cl* centre left; *cr* centre
right; *bl* bottom left; *bc* bottom centre; *br* bottom right.

BBC Hulton Picture Library: pp 42*bl*, 69, 88*b*, 93, 96, 98*t*,
103*b*, 112*t*, 115, 117, 135, 147, 157, 158*tr*, 167, 172*b*, 178, 181*b*.
Bodleian Library, Oxford: Filmstrip 153A-pp 2*l*, 27, 37,
59*tr*, 71*t*, 79, 134, 141*b*, 143, 163. Filmstrip 170H-pg 103*t*.
Gene Cox: pg 122*t*.
Mary Evans Picture Library: pp 1, 10*t*, 13, 17*b*, 18*tl*, 25*t*,
61*tr*, 73, 75, 89*c*, 100*b*, 100*tc*, 101*cr*, 106*b*, 107*b*, 108*b*, 120, 123,
126*b*, 127*tl*, 138*t*, 146*c*, 156, 180.
The Mansell Collection: Cover *tl*, pp 2*r*, 8, 9, 11*t*, 11*b*, 12,
14*bl*, 42*tl*, 45*b*, 49, 54, 57*t*, 67*br*, 72*t*, 72*b*, 74*c*, 84, 87, 91*b*, 100*t*,
100*c*, 101*b*, 106*t*, 107*t*, 107*t*, 109*b*, 124*t*, 132*b*, 145*t*, 145*c*, 154*t*,
160*b*, 161*t*, 169, 170/171, 172*t*, 173.
Civic Museum at Padua: pg 19.
Barry Richards: pg 47.
Ann Ronan Picture Library: pp 4*c*, 16, 17*t*, 21, 53, 55, 64,
65, 83, 90*br*, 97, 101*cl*, 101*c*, 104, 105, 108*t*, 110, 121*t*, 126*t*,
127*tr*, 127*bl*, 139*bl*, 139*br*, 154*b*, 158*br*, 177, 179.
**Ann Ronan Picture Library and E. P. Goldschmidt &
Co. Ltd.:** pp 15, 76, 80*b*, 176.
By Kind Permission of the President and the Council

of the Royal College of Surgeons of England: pp 24*l*,
40*tr*, 42*tr*, 42*br*, 43, 61*ct*, 80*t*, 90*bl*, 99, 152*l*, 152*r*, 155.
Scala/Vision International: Cover *bl*, pg 33.
Science Photo Library/Ohio Nuclear Inc.: Cover *tr*,
pg 95.
Ronald Sheridan's Photo Library: pp 20*tl*, 32.
C. James Webb: pg 122*b*.
By courtesy of the Wellcome Trustees: pp 2*c*, 3*r*, 4*l*, 4*r*,
14*tr*, 25*c*, 29*tl*, 29*c*, 29*b*, 41*c*, 41*b*, 44, 45*c*, 46, 52*t*, 52*c*, 52*b*, 56,
57*b*, 60*bl*, 62, 65*br*, 66*tl*, 66*bl*, 74*b*, 88*t*, 89*t*, 91*t*, 92*t*, 92*b*, 98*cl*,
112*b*, 124*b*, 129, 130, 131, 132*t*, 133, 138*b*, 142, 145*b*, 148*b*.
**Professor M. H. F. Wilkins, King's College
Biophysics Dept., London:** pg 125.
**Royal Library, Windsor Castle—Reproduced by
Gracious Permission of Her Majesty the Queen:** Cover
br, pp 3*l*, 22, 30, 34*tl*, 39, 51, 68*b*, 150, 153.
Carl Zeiss (Oberkochen) Ltd.: pg 127*br*.

In addition the publishers particularly wish to thank
Routledge & Kegan Paul for the use of material from the
book *The Evolution of Anatomy* by Charles Singer pub. 1925;
Anne Barrett for the special illustrations which appear on
pages 20, 26, 34, 35, 48, 58, 70, 73, 76, 77, 78, 96, 102, 110,
111, 118, 128, 134, 140, 146, 150;
Doreen Blake, who prepared the index.